NATURAL DISASTERS
Policy Issues and Mitigation Strategies

NATURAL DISASTERS
Policy Issues and Mitigation Strategies

- Editor -
Andi Eka Sakya

CENTRE FOR SCIENCE & TECHNOLOGY OF THE NON-ALIGNED AND OTHER DEVELOPING COUNTRIES (NAM S&T CENTRE)

2011
DAYA PUBLISHING HOUSE®
Delhi - 110 002

Centre for Science and Technology of the Non-Aligned and Other Developing Countries (NAM S&T Centre)
Core-6A, 2nd Floor, India Habitat Centre, Lodhi Road,
New Delhi-110 003 (India)
Phone: +91-11-24644974, 24645134, Fax: +91-11-24644973
E-mail: namstct@gmail.com
Website: www.namstct.org

Published by : **Daya Publishing House®**
A Division of
Astral International Pvt. Ltd.
– ISO 9001:2008 Certified Company –
4760-61/23, Ansari Road, Darya Ganj
New Delhi-110 002
Ph. 011-43549197, 23278134
E-mail: info@astralint.com
Website: www.astralint.com

Laser Typesetting : **Classic Computer Services**
Delhi - 110 035

Printed at : **Chawla Offset Printers**
Delhi - 110 052

PRINTED IN INDIA

Foreword

Natural disasters, regardless of country or territory or time they strike, leave behind a mind-boggling trail of their signatures and countless eye-opening lessons no university of the world can ever teach. The conventional university education teaches us first, and takes our test afterwards. Disasters do exactly the opposite. They take our test before they teach us in the live laboratory of Nature. It is therefore, only wise that we beat the test-teach cycle of the Nature by learning from the past disasters, if we are really serious about reducing their negative impact until we perfect the art of disaster prevention.

The learning process, like the Vedas, is without a beginning or without an end and invariably throws us into the vortex of the paradox that more we learn, the more remains there to be learned. When exposed to the aftermath of a severe natural disaster anywhere on the globe, the sky becomes the limit to which the learning process could, and should go on. And there is enough to learn, for every country, at all levels of hierarchy, at all times. We must begin with a vow to learn from every disaster, not to repeat the past mistakes, convert every disaster into a great learning opportunity and declare a collective war against our own exploitative tendencies and violence against nature that lie at the root of the problem.

Natural Disasters Mitigation being a subject of profound global concern, hundreds of teams and taskforces across the globe are perennially busy addressing the multifaceted aspects of natural disasters in the different geo-climatic, cultural and socio-economic settings. Since developing countries are the worst hit by natural disasters and their problems are more or less of the same genre, building joint programmes and partnerships in science and technology seems quite natural these days. Since most of the developing countries lack even the minimal of resources and wherewithal to fight natural disasters, they regard the hand holding exercise as a very wise move of great potential. And indeed wherever win-win partnerships have emerged, the cause of natural disaster reduction has flourished, without question.

On the other hand, wherever developing countries have tried to stand in isolation either by accident or by the design, the progress has been tardy.

The International Roundtable on Natural Disaster Management: Lessons from Natural Disasters, Policy Issues and Mitigation Strategies held during 8-12 January, 2007 at the joint initiative of the Centre for Science and Technology of the Non-aligned and Other Developing Countries (NAM S&T Centre) and Centre of Disaster Mitigation and Management (CDMM) of the Vellore Institute of Technology University was expressly designed to promote such partnerships. It may also be seen as a pace-setter initiative by the NAM S&T Centre to create an interactive platform for exchange of experiences on disasters. The two visible gains of the initiative are the Vellore Declaration 2007 that sets the agenda for action and this compendium of country reports now in your hands that walks you through the proceedings of the roundtable. It also has potential to lead to bilateral and multi-lateral programmes between and among the participating member States.

The fate of millions of people in the world is written and rewritten, and often underscored, not in the texts of the developmental plans and reports but by the onslaught of frequent cataclysmic events. Disasters often lead and the Plans follow. Survivors are condemned to their fates worse than dead. Besides huge losses of lives and property, the very wheels of progress and development get jammed. Sometimes, even if the initial impact of a disaster is not quite severe, its eventual consequences could still be lethal. For example rainstorm–flood contamination–epidemics; or cyclone–flood contamination–epidemics can end up with a mass graveyard scenario, throwing even the best of the post disaster management machinery haywire.

Let us resolve not to miss any opportunity any more and learn to fight disasters together in a spirit of co-operative endeavour. There are positive sides of disaster as well. Disasters help us to test our ideas on the whole range of issues, as also give us the opportunity to learn and innovate. Every disaster provides policy makers the heaven sent opportunity to put their policies on the anvil. Professionals charged with the responsibility to manage natural disasters, likewise, get a good opportunity to introspect and see by hindsight where their preparedness plans and strategies let them down and why? Scientists do get a load of new problems, enough food for thought as also fresh ideas and the rare ammunition to re-write their research proposals in search of cost effective solutions. Since dead tell no tale, those who survive are our best informers.

When a disaster strikes, although we cannot recreate the lost lives, we can certainly endeavour to ensure that no more lives will be lost on repeat of such events in future by converting every disaster into an opportunity to learn and to build back better. For example, we can resolve that unplanned, non-engineered constructions razed to ground by disasters will be replaced by disaster resistant, construction engineered to take into account the future needs. When planning for immediate action, let us use all our collective wisdom, education, training and tools, to quickly respond to every disaster. When planning for a medium term of a decade or so, let us aim at enhanced safety through a systematic plan. And when planning for the long-term future, let us educate people about the great culture of safety which is our ultimate goal. Let us not forget that all the thinking (knowledge) and good work (action) are required to be done between the two disasters. Thomas Henry Huxley once said that the great end of life is not knowledge but action.

If this event of roundtable will generate some action towards ushering the great culture of safety, the temerity of my association as the coordinator of the event will become a matter of lasting personal joy.

New Delhi: February 2011

R.K. Bhandari
Founder Chairman
Centre for Disaster Mitigation and Management
VIT University, Tamil Nadu, India

Preface

Natural disaster is unquestionably one of the most frequent issues heard recently. The natural disasters by which the least developed and developing countries suffer most are not only caused by the recent concern of climate change, but also caused by their geographic position naturally pose the hazards. The raising concerned on natural disasters have made many countries – moreover the least developed and developing ones and especially the most vulnerable ones – taking precautionary measures to anticipate the impact.

Either geophysical related disasters such as earthquake, tsunami, high wave or climate related ones, such as drought and flooding, cause human sadness and desolation, as well as exacerbate conflict over limited resources. Moreover, the problems are worsened by the fact that disasters could only be solved thoroughly, if ones were comprehensible to several aspects, including – among others – the cause and warning (mitigation), and the evacuation and recovery (adaptation). Regardless of the aspects, the significance of the problems necessitate that one's act now assuming the disasters tends to be getting frequent and aggravated.

Realizing the urgent predicament, the Centre for Science and Technology of the Non-aligned and other Developing Countries (NAM S&T Centre) in India organized a 5-days international roundtable on Natural Disaster Management with the theme of Lessons from Natural Disasters, Policy Issues and Mitigation Strategies at Vellore, India during 8–12 January, 2007 in association with the Centre for Disaster Mitigation and Management (CDMM) of the VIT University. Twenty eight participants of the Roundtable from 13 countries included 13 overseas experts and professionals from around the world, ranging from South Africa to Gogota in South America.

Managing disaster mitigation and adaptation – although it is quite an old issues – is indeed still relatively a new topic from the perspective of systematic and wholeness solution approach. There might be that every country has its own way of tackling the problem depending on their intrinsic position. Therefore, learning from other countries successfully solving their problem can serve the best and the quickest way in comprehending the solution. Realizing that the immense difficulties cannot be borne and solved by a single country, the participants signed the so-called Vellore Resolution at the final day of the discussion. The resolution called for the urgent steps of raising-up the disaster awareness from every organization and country right from the planning phase.

This book is a collection of all country reports as well as expert views participated in the meeting. The aim of S&T NAM Centre is to facilitate strategic and policy oriented assessment by scientifically concised methods and analysis. By having this collected reports, it is believed that not only are these objectives fulfilled, but also does give a mean to defining appropriate measures for anticipating the increasing frequency of the natural disasters occurrence.

Andi Eka Sakya
Indonesia

Introduction

A disaster is characterised by an occurrence such as tsunami, earthquake, flood, fire, hurricane, tornado, storm, drought, explosion, volcanic eruption, spread of epidemic, building collapse, transportation accident or any other situation leading to the human suffering, or creating such situation in which the victims needs assistance to alleviate the suffering. Any disaster significantly impacts the economic growth and development, social and economic infrastructure and the environment. In many cases, a recurring or forceful disaster may keep a developing country in severe perennial poverty. Risk assessment; forecasting, monitoring and early warning; emergency management; developing a disaster prevention strategy; improving awareness; integrated sustainable development; and also a political will and commitment with the involvement of professional and scientific bodies and public-private partnership are some of the key elements in context with the natural disasters. The vulnerability to natural disasters combined with socio-economic vulnerability of the people poses a great challenge for the government machineries and underscores the need for comprehensive plans for disaster preparedness and mitigation, as well as the training and capacity building of the officials dealing with emergency situations. Every disaster provides policy makers an opportunity to put their policies on the anvil. Likewise, the professionals charged with the responsibility to counter natural disasters get a good opportunity to introspect and see by hindsight where their preparedness plans and strategies have failed them, and why? Scientists too get their food for thought as also fresh ideas and the rare ammunition to re-write their research proposals towards the fulfilment of their insatiable quest for improved, cost effective solutions. Since the hazardous events are frequent, degree of preparedness is often low and since the urbanisation is out of control, the losses are staggering. The strategic thinking is needed to unfold scenarios before they really occur so that road map could be updated, game plan could be revised and strategic sense could be sharpened. Sharing of resources, pooling of expertise and leveraging of capacities come naturally with such planned thinking. Since developing countries are the worst hit by natural disasters and lack even the minimal of resources and wherewithal to fight natural disasters, their problems are more or less of the same genre. Therefore it is essential to build joint programmes and win-win partnerships on natural disasters management and mitigation.

With the aforementioned in view, the NAM S&T Centre organised a 5-days international roundtable on Natural Disaster Management with the theme of 'Lessons from Natural Disasters, Policy Issues and Mitigation Strategies' at Vellore, India during 8–12 January, 2007 in association with the Centre for Disaster Mitigation and Management (CDMM) of the VIT University. Twenty eight participants of the Roundtable included the experts and professionals from 13 countries, namely, Bangladesh, Brunei Darussalam, Colombia, Indonesia, Malaysia, Mauritius, Myanmar, Pakistan, South Africa, Sri Lanka, United Arab Emirates and Vietnam and the host country India. The lectures on various facets of disaster management and mitigation, such as earthquakes, landslides, wind disasters, cyclones, floods and tsunamis were delivered by eminent Indian scholars. The overseas participants made power point presentation of the country status reports and in some cases, showed short films on the disasters and mitigation and rehabilitation efforts. All the roundtable participants also undertook three Auto Certification Tests, respectively on Cyclones, Earthquakes and Landslides, for which the CDs were prepared by CDMM. The Roundtable concluded with a panel discussion on disaster management policies and adoption of a 'Vellore Resolution 2007'.

The present publication is based on the presentations made by the participants of the Vellore Roundtable and also includes some papers contributed by other eminent specialists, and includes 16 research papers and country status reports from 12 developing countries of Asian, African and Latin American region.

I gratefully acknowledge the dynamic involvement and untiring efforts of Dr. Andi Eka Sakya, Executive Secretary, Agency for Meteorology, Climatology and Geophysics of Indonesia (BMKG) for technical editing of this publication. I am indebted to Prof. R.K. Bhandari for sparing his valuable time in writing a 'Foreword' for this book. Last, but not the least, the valuable services rendered by the entire team of the NAM S&T Centre under the guidance of Mr. M. Bandyopadhyay and Dr. V.P. Kharbanda, particularly Mr. Gaurav Gaur, Mr. Yasir Abbas Rizvi and Mr. Pankaj Buttan, in compiling the presented papers and giving a shape to this volume are deeply appreciated.

I hope that this publication will serve as a valuable reference material for the researchers and professional engaged with the disaster management issues.

Arun P. Kulshreshtha
Director, NAM S&T Centre

Contents

Chapter 1

ICSU ROA's Science Plan to Address Natural and Human-Induced Environmental Hazards and Disasters in Sub-Saharan Africa

Genene Mulugeta[1], Ray Durrheim[2*], Samuel Ayonghe[3], Deolall Daby[4], Opha Pauline Dube[5], Francis Gudyanga[6] and Filipe Lucio[7]

[1]Uppsala University, Sweden
[2]CSIR and University of the Witwatersrand, South Africa
[3]Department of Geology and Environmental Science, University of Buea, Cameroon
[4]Department of Biosciences, University of Mauritius
[5]Department of Environmental Sciences, University of Botswana
[6]Research Council of Zimbabwe
[7]National Institute of Meteorology, Mozambique
*E-mail: rdurrhei@csir.co.za

ABSTRACT

The authors identify several factors that contribute to Africa's high vulnerability to disasters, including the high rate of population growth, food insecurity, high levels of poverty, inappropriate use of natural resources, and failures of policy and institutional frameworks. Despite the huge negative impact that natural and man-made disasters make on Africa's development, little is done to prevent them. Effective strategies to prevent hazards becoming disasters and to manage those disasters that do occur would make a lasting contribution to the quality of life and sustainable livelihoods of Africans.

The ICSU Regional Office for Africa (ICSU ROA) Scoping Group on Natural and Human-Induced Environmental Hazards and Disasters proposes the establishment of a research, capacity building and outreach programme aimed at reducing the risk of disasters and increasing resilience. The main focus of the programme is the development of a truly regional and inter-disciplinary approach to the understanding, prediction, assessment and mitigation of hazards and disasters.

ICSU ROA offers the opportunity to bring together existing institutions, appropriate partners (such as universities, scientific institutions, development agencies, humanitarian assistance agencies and NGOs), and policy makers to further develop and build on the activities identified in this strategy. Details of how ICSU ROA intends to achieve these objectives are outlined in the science plan.

Introduction

Africa is a continent prone to a wide variety of natural and human-induced environmental hazards and disasters. Phenomena such as floods, hurricanes, earthquakes, tsunamis, drought, wildfires, pest plagues, air, water and soil pollution cause extensive losses to livelihoods and property, and claim many lives. The population of Africa, estimated at 880 million in 2005, is increasing at a rate of 2-4 per cent per annum, so the number of people exposed to environmental hazards and disasters will continue to increase. However, mitigation measures are a relatively low priority for African decision and policy makers, as environmental hazards and disasters often pale into insignificance when compared to other pressing issues such as poverty and HIV/AIDS. The fact that 43 African countries are heavily indebted (OFDA-CRED, 2002; UN-ISDR, 2004), makes Africa the least equipped and prepared continent to cope with the impacts of hazards and disasters. The reduction of disaster risk through preventive measures is thus a central concern for the sustainable development of Africa. It is vitally important that African countries adopt cost effective policies to lower risk and allocate appropriate resources for hazard and disaster mitigation.

Africa is, in many ways, the continent that is most in need of scientific knowledge to provide solutions to its socio-economic development. However, the latest developments in science are not often readily available to scientists in Africa. At the same time, investment in science is frequently a low priority for Africa decision and policy-makers, and many scientific institutions have relatively weak infrastructures. Globally, most current research is directed towards the North and its problems, while the significant societal problems of the South are largely unaddressed.

In an initiative launched in 2006, the International Council for Science Regional Office for Africa (ICSU ROA) seeks to revitalise efforts to address the impact of environmental hazards on African communities. It is a major challenge for the African scientific community to develop a truly regional and global partnership to minimise these impacts. ICSU ROA's overall objective is to contribute to improved risk management and to assist in building a culture of prevention by improving public awareness and facilitating accessibility to disaster information through joint initiatives with other national, regional and international organisations, governments and civil societies, for the sake of sustainable development of Africa.

ICSU ROA seeks to develop both long- and short-term action plans to implement the strategy. At present, disaster management in Africa is largely limited to emergency humanitarian assistance. In the long term, ICSU ROA aims to mainstream disaster risk reduction practices into knowledge management in order to reduce vulnerability to future hazards and disasters. Moreover, ICSU ROA will work towards advocacy for incorporation of research findings into policies, and will facilitate planning guides and training activities at all levels in society. However, this requires a multi-disciplinary approach to overcome the limited capacity of scientific research institutions in Africa. Short-term activities included the production of a science plan that was approved by the ICSU Regional Committee for Africa at its meeting in the Seychelles in March 2007 (Mulugeta *et al.*, 2007), participation in the hazard-related activities of the International Year of Planet Earth (IYPE) in 2007-2009, and the production of a book on "the societal impact of natural and human-induced hazards and disasters in Africa"

Environmental Hazards and Disasters in Sub-Saharan Africa

A hazard is any event, phenomenon, or human activity that may cause loss. Natural and human-induced factors may act together to create a hazard. For example, earthquakes are usually considered to be natural hazards, but they may also be triggered by mining activities or the impoundment of large dams. A landslide may be caused by a combination of heavy rains, light earth tremors, and deforestation. A disaster is an event that causes a serious disruption, leading to widespread human, material or economic losses beyond the coping capacity of a given society. Disaster management requires a set of actions and processes designed to lessen hazardous events before they become disasters.

The Earth Institute at Columbia University (USA) conducted a project assessing natural disasters and risks to human populations and economic activity to provide a quantitative basis for risk-conscious investments in sustainable development worldwide (Dilley *et al.*, 2005). The study compiled event data for various natural hazards (Figure 1.1).

The report notes, "drought and combinations of drought and hydro-meteorological hazards dominate both mortality and economic losses in sub-Saharan Africa". In no other continent does drought appear to be as severe a

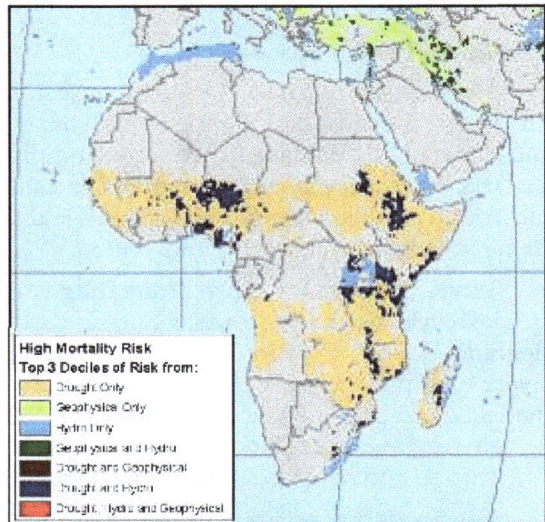

High Mortality Risk
Top 3 Deciles of Risk from:
- Drought Only
- Geophysical Only
- Hydro Only
- Geophysical and Hydro
- Drought and Geophysical
- Drought and Hydro
- Drought, Hydro and Geophysical

Figure 1.1: African Natural Disaster Hotspots[4].

risk as in Africa. While some developed countries are regularly threatened by sudden and dramatic events such as hurricanes, floods and earthquakes, there is great awareness of these hazards and the need to prepare for them. In developed countries, estimates of losses usually reflect insured losses of physical infrastructure in densely populated areas. In contrast, most hazards and disasters in Africa (with a few exceptions such as the Mozambique floods of 1999/2000) are relatively silent and insidious encroachments on life and livelihood that increase social, economic, and environmental vulnerability to even modest events. For example, recurrent drought, deforestation and progressive land degradation, desertification, and HIV/AIDS result in incalculable human, crop, livestock, and environmental losses which are not easily measured by conventional disaster-loss tracking systems (Holloway, 1999; Swiss Agency for Development and Cooperation, 2006). As a result, the losses caused by African disasters are often under-estimated.

Hydro-meteorological Hazards

Hydro-meteorological events give rise to the majority of disasters in Sub-Saharan Africa, impacting nearly every country. These include floods, tropical cyclones, storm wave surges, droughts, extreme temperatures, forest/scrub fires, sand or dust storms, landslides and avalanches. In the period 1975-2002, disasters of hydro-meteorological origin constituted 59 per cent of the total number of natural disasters in Sub-Saharan Africa, with floods accounting for 27 per cent, droughts for 21 per cent, windstorms (particularly tropical cyclones) for 9 per cent, and wildfires for 1 per cent (OFDA-CRED, 2002; UN-ISDR, 2004). An alarming trend is the increasing number of people affected by natural hazards of hydro-meteorological origin, with drought, flooding and windstorms accounting for 90 per cent of the total number of people affected. Global climate change will continue to alter the risk associated with hydro-meteorological hazards. The vulnerability of Africa's environment is exacerbated by land degradation, which is a major environmental hazard on the continent.

Floods and Flash Floods

Floods, including flash floods that arise from tropical cyclones and severe storms, are among the most devastating natural hazards in Africa. Floods and flash floods cause loss of life, damage to property, and promote the spread of diseases such as malaria, dengue fever, cholera, typhoid and chikungunya. From 1900 to 2006, floods in Africa have killed nearly 20,000 people and affected nearly 40 million, and caused damage estimated at nearly US$ 4 billion.

While the primary cause of flooding is abnormally high rainfall (*e.g.* due to tropical cyclones), there are many human-induced contributory causes such as: land degradation; deforestation of catchment areas; increased population density along riverbanks; poor land use planning, zoning, and control of flood plain development; inadequate drainage, particularly in cities; and inadequate management of discharges from river reservoirs. Flooding can also be caused by the failure of dams, both constructed and natural (*e.g.* the pyroclastic dam at Lake Nyos in Cameroon).

The floods that occurred in Mozambique in February 2000 are a recent example of a flood disaster. Rainfall accompanying tropical cyclone Eline caused excessive flows in rivers such as the Limpopo River with catchments in other countries. These floods affected a total of about 4.5 million people, caused 700 deaths, losses estimated at US$ 500 million, and the GDP growth rate decreased from 10 per cent to 2 per cent . Between 2005 and 2007, flooding afflicted several areas in eastern and southern Ethiopia, Somalia and Kenya, killing and displacing hundreds of people. The Shabelle and Juba rivers in the region have both flooded their banks, affecting towns and villages in an area stretching hundreds of kilometres. Floods in the Horn of Africa normally follow the June-September rainy season. According to the UN, the 2006 floods, which followed droughts in 2005, affected 1.8 million people and were the worst in the region for 50 years. In August 2006, overflow of the Dechatu River killed more than 300 people in Dire Dawa (a town in south-eastern Ethiopia), displaced thousands more, and caused extensive damage to homes and markets (Scores killed in Ethiopia floods, BBC News, 6 August, 2006).

Flood defence is essential to protect communities. Self help for long-term mitigation should be encouraged. At present the accuracy and lead times of flood forecasts in sub-Saharan Africa are limited or questionable. Thus, training and research should stress the prevention of floods. New research and collaborative efforts are needed to advance flood management in the future.

Droughts

Future projections show a net overall global drying trend and the proportion of the land surface in extreme drought is predicted to increase from 1 per cent for the present-day to 30 per cent by the end of the 21st century. The drying trend is related to anthropogenic emissions of greenhouse gasses and sulphate aerosols into the atmosphere (Burke *et al.*, 2006). Although droughts currently affect many parts of the globe, they are a particular concern in sub-Saharan Africa. Emergency food aid to Africa currently accounts for around 50 per cent of the budget of the World Food Aid Programme (Economist, 2006).

A large part of the Sub-Sahara is susceptible to drought, especially in the Sahel with annual rainfall of 150-600 mm, while much of southern Africa, including regions outside the Kalahari, experience frequent drought. The Sahel experienced devastating and prolonged droughts that lasted up to 30 years starting from the late 1960s, the causes of which remain a subject of debate. Initial studies blamed the persistence of the drought on poor land use and the resulting desertification, but recent work indicates that the 3-decade-long drought might have been due to complex interactions among the atmosphere, land, and ocean (Nicholson, 2005; Foley *et al.*, 2003). The 1970-1974 droughts in the Sahelian region caused unprecedented losses in human life, livestock and environmental damage. The drought was equally devastating in the Horn of Africa, and Ethiopia suffered heavily with an estimated 250,000 lives and 50 per cent of livestock lost in the Tigray and Welo regions. The widespread droughts of 1984-1985 were the most significant: about 8 million people were affected, 1 million died, and large numbers of livestock were lost in the Horn of Africa (Webb *et al.*, 1991). In the 2000 drought, nearly 100,000 people died in the same region. The most severely affected were the 16 million nomadic pastoralists whose range straddles the borderlands between Kenya, Somalia and Ethiopia (UNICEF, 2006). In 2006, it was reported that over 8 million people were on the brink of starvation in the Horn of Africa (Kenya, Djibouti, Ethiopia, Eritrea and Somalia) due to severe drought, crop failure and loss of livestock (UNICEF, 2006).

Remote sensing data indicated a degree of vegetation recovery in the Sahel from the late 1980s to 2003, although most rainfall records were below the long-term mean (Nicholson, 2005; Anyamba and Tucker, 2005). The cause of the green-up is still debated. For example, Herrman *et al.* (2005) suggest that CO_2 fertilisation and the resulting increase in the efficiency of water use by plants could play a role.

In southern Africa, severe droughts (such as those of 1982–1983 and 1997–1998) have been linked to the *El Nino*-Southern Oscillation (ENSO) phenomenon. Nearly all climate change projections signal greater chances of severe droughts over southern Africa, particularly the central to western areas (IPCC, 2001; Scholes and Biggs, 2004).

Drought is also exacerbated by deforestation. For example, deforestation rates in the Congo Basin rainforest were estimated at 0.6 per cent per year in the period 1980–1990, while rates for the whole of Africa vary from 0.1 to 0.7 per cent (Scholes and Biggs, 2004). Deforestation leads to land degradation and eventually desertification, thus increasing the vulnerability of populations to drought (Timberlake, 1994). The most serious result of drought is famine. However, drought and famine are not sudden events, but rather the end result of long-term degradation of the environment due to poor land use and deforestation, as well as other factors such as poor distribution of food.

There are a number of organisations that operate to combat drought in sub-Saharan Africa. The Economic Community of West African States (ECOWAS) is developing programmes in environment and natural resource management, including desertification and water control management. The South African Development Community (SADC), through its SADC Water Sector coordinating unit, has approved a strategic approach to manage drought and floods. The key institutional player is the SADC Drought Monitoring Centre, based in Gaborone, Botswana. The Regional Early Warning Unit (REWU) develops information on weather threats, conditions and drought, and works closely with the African Centre of Meteorological Application for Development (ACMAD) in Niger. ACMAD's mission is to provide weather and climate information to member countries through weather prediction, climate monitoring, technology transfer (telecommunications, computing and rural communication) and research. The Inter-Governmental Authority on Development (IGAD) operates a Regional Early Warning System (REWS) as a key component of national drought and flood preparedness in the Horn of Africa. The IGAD Climate Prediction and Applications Centre (ICPAC) in Nairobi, Kenya, is responsible for the Great Horn of Africa Climate Outlook Forum (GACOF), a participatory consensus mechanism for deriving seasonal forecasts. In the SADC region the same process is known as SARCOF. The AGRHRYMET Regional Centre in Niger is a specialised centre for training and applications in agro-meteorology and operational hydrology. However, the services of these few technical institutions are limited due to lack of resources and capacity.

Heat Waves

Climate change studies show that Africa, like the rest of the world, became warmer over the past century (Figure 1.2). Temperatures are expected to continue to rise in the future. Extreme events such as heat waves are predicated to be one of the hazards that will be associated with climate change (Easterling *et al.*, 2000). Climate change studies focusing on trends on heat waves in Africa are lacking. However, indications for other parts of the world such as North America and Europe are that global warming will lead to more intense, frequent, and longer lasting heat waves over the 21st century (Meehl and Tebaldi, 2004).

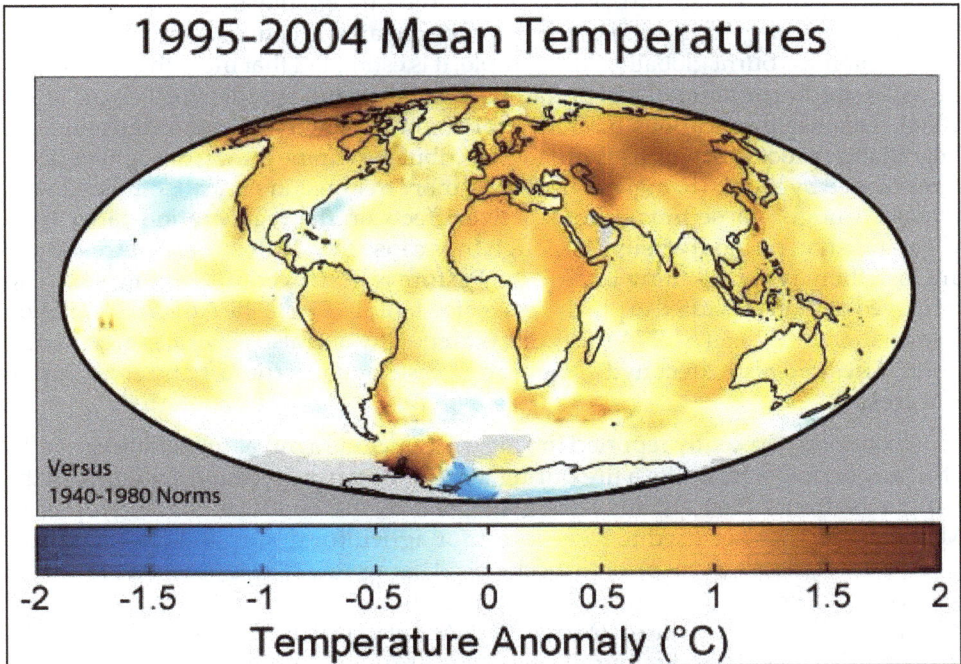

Figure 1.2: Global Mean Temperature Anomaly, 1995-2004 (IPCC, 2001).

Depending on the average weather in a particular area, a heat wave is a prolonged period of excessively hot weather, which may be accompanied by high humidity. There is no universal definition of a heat wave (Robinson, 2001) and there is generally little reporting on the health effects of extraordinarily hot conditions particularly in Africa. In Europe and North America, heat waves are increasingly contributing to weather-related deaths, as severe heat waves can lead to deaths from heat stroke. The elderly, very young children and those who are sick or overweight are at a higher risk of heat-related illness (Díaz *et al.*, 2004). Increased demand for cooling in cities during a heat wave often leads to electricity spikes that can create power outages, further exacerbating the problem. Heat waves may also lead to damage on infrastructure, for example, ruptured water lines and buckled roads. When a heat wave occurs during a drought, it can trigger hazards such as bushfires that threaten livelihood systems and security of people and animals.

The problem of hazards such as heat waves in Africa will be exacerbated by changes in lifestyle linked to urbanisation and general lack of preparation for such events. Moreover, the problem will be exacerbated by changes in lifestyle linked to urbanisation.

Fires

Much of Sub-Saharan Africa is susceptible to wildfires that destroy pastures, crops, buildings and infrastructure. Wildfires may be ignited naturally by lightning or the spontaneous combustion of coal (Zimbabwe) and peat (Okavango delta and Lesotho highlands). However, human beings are responsible for igniting most wildfires. About 168 million hectares burn annually south of the equator, accounting for 37 per cent of the dry biomass burnt globally. For example, it is estimated that more than 60 million hectares are burnt annually in Sudan (Goldammer and Ronde, 2004; Pyne *et al.,* 2004). This has implications on both short-term productivity and long-term land degradation processes, which eventually contribute to famine during drought periods. Fires caused by human beings are becoming more frequent in Africa. Combined with intense drought, these fires have negative effects on the regeneration capacity of vegetation and on biodiversity. Although fires cause few deaths, valuable resources are lost, thereby contributing to poverty. Pasture is destroyed, and animals have to be moved or funds allocated to purchase their feed. According to the Africa Air Pollution Information Network (APINA), fire affects air quality and also generates greenhouse gases. Fires can also affect hydrological processes such as run-off and may lead to soil erosion.

Examples of recently reported fire incidents include a threat to fuel storage tanks at an airport in Botswana by a fire during the dry season of 2005; and a wildfire in the Kruger National Park, South Africa, that led to the death of ca. 20 people. In Madagascar, fires are used to clear forest for agricultural purposes. In Mauritius, sugar cane fields are burnt prior to mechanised harvesting. These fires sometimes get out of control, causing ecological disaster. In the Borana Zone in Ethiopia, over 32,000 hectares were burnt by 96 fires in March 2000. About 80,000 fire fighters were mobilised to extinguish these fires (Freiburg University, 2006).

Fire prevention and mitigation requires knowledge on weather, ecology and terrain of the area; infrastructure such as machinery; use of satellite images for monitoring; ability to mobilise and train human resources; and the availability of communication and road networks, all of which are scarce in Sub-Saharan Africa. A few African countries, such as Ethiopia and South Africa, have fire danger warning systems. However, most research is based on ecological field studies. Recently, the use of satellite data to monitor burnt areas for purposes of estimating biomass-related greenhouses gases has been introduced. The Southern Africa Fire Network (SAFNet) provides a framework for the exchange of fire management information and capacity building, with the emphasis on the use of geo-spatial information technologies. The Global Fire Monitoring Centre (GFMC) also covers fires in Africa. In addition, the University of Maryland (USA) and NASA provide near real time information on active fires detected by the MODIS satellite (NASA, 2006). In Southern Africa this service is provided through collaboration with the South African Satellite Application Centre (SAC).

⌐pical Cyclones and Hurricanes

Weather systems characterised by extreme winds and rainfall, known as tropical cyclones in the Indian Ocean and hurricanes in the Atlantic Ocean, are generated between latitudes 5° to 20° when sea temperatures are sufficiently warm. Cyclones and hurricanes are capable of annihilating coastal areas with sustained winds of 250 km/h or higher, heavy rainfall, and, most devastatingly, storm surges that cause the ocean level to rise by as much as 10 metres. As a cyclone approaches the coast, an 80 to 160 km diameter dome of ocean water sweeps over the coastline, causing coastal flooding and damage to coral reefs, mangroves and fisheries. In most low income countries the mortality rates associated with cyclones are 3 to 20 times larger than those associated with floods. Tropical cyclones can cause huge economic losses, especially on island states, by damaging dwellings, infrastructure (power, telecommunications, roads) and fisheries. Heavy rainfall can cause floods that damage infrastructure and crops, trigger landslides, and promote disease. The impact of these storms on coastal communities is exacerbated by the destruction of natural barriers such as mangrove swamps.

In Sub-Saharan Africa, the areas most frequently affected by cyclones are the Indian Ocean islands and the coastal areas of eastern and southern Africa. Cyclones can penetrate inland as far as Botswana. Countries such as Mauritius are well prepared for cyclones, while countries such as Madagascar, Comoros, and Mozambique are more vulnerable to cyclones. Typically, 12 cyclones occur annually in the southwestern Indian Ocean. A very severe cyclone occurs every 10 years on average. There is concern that Atlantic Ocean hurricanes could affect West African countries such as Senegal, although there has been no recorded instance of this happening. Further research is needed to assess the risk.

The WMO Regional Specialised Meteorological Centre in Reunion serves the sub-region with information concerning cyclone disasters, especially the members of the South West Indian Cyclone Committee (SWIO). Cyclone warnings are broadcast on radio and television, and published in the press. Warnings are also disseminated through local structures such as schools, religious networks and government and traditional structures. This has made it possible for countries such as Mauritius to reduce the number of people killed by cyclones.

Severe Storms

Tornados are violent rotating columns of air extending from thunderstorms and are amongst the most violent and destructive of all weather phenomena. Hailstorms are associated with thunderstorm activity caused by intense convection and occur in areas such as the South African highveld, causing damage to property, crops and livestock. The forecasting of tornados and hailstorms is very challenging as they affect highly localised areas and last on average less than 30 minutes. Currently, there are no warning systems for tornados and hailstorms anywhere in Africa. Losses due to hailstorms and tornados in Africa are limited compared to other environmental hazards, and little research has been conducted.

Dust Storms

The Sahel region is one of the largest sources of dust storms in the world. Summer storms are due to gusts associated with convective rain-bearing storm systems, while winter storms are associated with the Harmattan winds. The dust alters the air quality, affecting animals, plants and the weather. Scientists in the Niger-based Centre de Recherche Médicale et Sanitaire (CERMES) have found that dust storms blowing across the Sahel might be linked to lethal meningitis outbreaks that often hit this region and its 300 million inhabitants.

Geological Hazards

Currently, disasters due to geological hazards have a far smaller impact on Sub-Saharan Africa than those due to hydro-meteorological hazards. Earthquakes account for 2 per cent , and landslides and volcanic hazards account for 1 per cent of the number of disasters on the continent (OFDA-CRED, 2002). However, the impact of these hazards may change in future.

Earthquakes

Sub-Saharan Africa is largely a stable intra-plate region characterised by a relatively low level of seismic activities, with earthquakes randomly distributed in space and time (Figure 1.3). The only parts of sub-Saharan Africa that do not display the characteristics of an intra-plate region are the East African Rift System and the Cameroon Volcanic Line, where earthquakes are associated with active fault zones and volcanic activities.

Damaging earthquakes with magnitudes greater than 6 occur almost annually in the East African Rift. Recent events include the February 2006 Mozambican M7.0 earthquake, which was one of the largest ever recorded in southern Africa. Shaking was felt as far as in Zimbabwe and South Africa. Four people were killed, 27 injured, and at least 160 buildings damaged. The extent of economic losses was not evaluated. In December 2005, an M6.8 event in the Democratic Republic of Congo (DRC) caused several deaths and damaged school buildings. The event also killed a number of people in the Lake Tanganyika region of western Tanzania, and left more than 400 families homeless.

The Cameroon Volcanic Line experiences earthquakes associated with volcanoes and fault movements. The earthquakes do not exceed magnitude 6, and so far have not caused any human casualties.

Earthquakes also occur occasionally in the Cape Fold Belt, South Africa. The most destructive earthquake that occurred in recorded history was a M6.3 event that took place on 29 September 1969 in the Ceres-Tulbagh region of the Western Cape, killing 12 people. Aftershock activity had virtually ceased, when an M5.7 event occurred on 14 April, 1970, causing further damage in the towns of Ceres and Wolseley.

The impoundment of reservoirs has also triggered earthquakes. For example, the filling of Lake Kariba and subsequent fluctuations of water level have been accompanied by seismicity, the largest event having M6.2. Seismicity has also been associated with the Gariep Dam in South Africa and the Katse Dam in Lesotho.

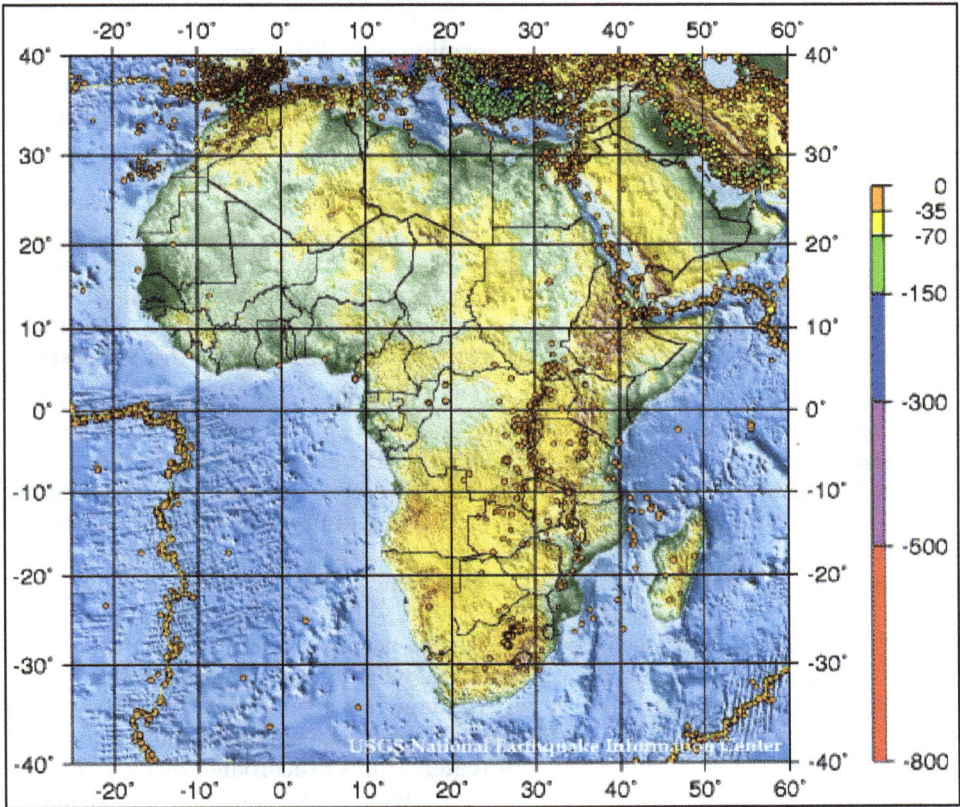

Figure 1.3: African Seismicity, 1990-2000 (Lockwood and Rubin, 1989).

Mining-related earthquakes pose a significant hazard to mineworkers in the gold and platinum mining districts of South Africa. Thousands of mineworkers have perished during the last century as a result of rock bursting. No member of the public has suffered fatal or even serious injuries due to mining-related earthquakes, although some events have caused damage to surface structures. For example, the M5.3 event that occurred on 9 March, 2005 near Stilfontein caused serious damage to schools, commercial properties, apartment blocks, the civic centre, and 25 houses.

The multitude of active faults in the East African Rift system poses a significant hazard. Several large dams have been built in the Rift system (*e.g.* Cahora Bassa, Kariba). However, African research institutions have limited capability to mitigate and respond to earthquake hazards. Currently, no earthquake warning in the region comes close to the required level of reliability. A sustainable earthquake disaster mitigation strategy requires compilation of base maps of known faults, as well efforts to detect possible unknown faults. It is also necessary to build interactive databases of high-risk areas and integrate these with population distribution, seismic history, and vulnerability to hazards and disasters. In order to advance seismic research, it is necessary to develop cooperation among existing institutions and networks such as

AfricaArray, a pan-African research and capacity-building programme launched in July 2004. AfricaArray is involved in determining the lithospheric structure of the African Plate, and the chemical and dynamic causes of the African Super Plume, the largest anomaly in the Earth's mantle, which occurs directly below South Africa.

Tsunamis

Tsunamis, also known as seismic sea waves, are a series of enormous waves created by an underwater disturbance such as an earthquake, landslide, volcanic eruption, or meteorite collision. A tsunami can move at hundreds of kilometres per hour in the open ocean and smash onto landmasses with waves as high as 30 metres or more. All tsunamis are potentially dangerous, even though they may not damage every coastline they strike. A tsunami can strike anywhere along the African coastline.

The 2004 Indian Ocean earthquake, known as the Sumatra-Andaman earthquake, was an undersea earthquake that occurred at 00:58:53 Coordinated Universal Time (UTC) (07:58:53 local time) on 26 December, 2004. The earthquake triggered a series of lethal tsunamis that spread throughout the Indian Ocean, killing large numbers of people and devastating coastal communities across South and South-East Asia, including parts of Indonesia, Sri Lanka, India, and Thailand. The number of casualties is estimated at 186,983 dead and 42,883 missing. The impact on coastal fishing communities has been devastating, with high losses of income earners as well as boats and fishing gear. Beyond the heavy toll on human lives, the tsunami caused an enormous environmental impact that will affect the region for many years to come. It has been reported that severe damage has been inflicted on ecosystems such as mangroves, coral reefs, forests, coastal wetlands, vegetation, sand dunes and rock formations, biodiversity and groundwater. This is exacerbated by the spread of solid and liquid waste and industrial chemicals, water pollution and the destruction of sewage collection and treatment systems. Soil and fresh water contamination with infiltrated salt water and salt layer deposits on arable land are also taking their toll.

Although not as severely affected as Asia, African countries also suffered losses (UNESCO-IOC, 2005). In Somalia, 176 people were killed, 136 went missing, and 50,000 were displaced. One person drowned in Kenya. In Madagascar, 1,000 people were left homeless. No casualties were reported in Mauritius, but a village in the northern island was submerged. Three people died and seven went missing in Seychelles. In Tanzania there were 10 deaths, an oil pipeline was destroyed and an oil tanker ran aground.

The disaster created an awareness of the need for a tsunami warning system for the Indian Ocean. Prior to the event of 26 December, 2004, very little research work had been done to address the risk of tsunamis in the region. No historical records of past tsunamis existed for the affected areas. A survey conducted by UNESCO/IOC, WMO and ISDR showed that African countries have very limited capacity to effectively implement mitigation measures for tsunamis (UNESCO-IOC, 2005). The United Nations started working on an Indian Ocean Tsunami Warning System, and by 2005 had the initial steps in place.

Volcanoes and Explosive Crater Lakes

Active volcanoes pose a serious threat to life and property in parts of Africa. Africa has about 140 volcanoes that have erupted during the last 10,000 years, of which 25 are active (*i.e.* have erupted during recent historic time, ca. 500 years). Volcanic eruptions produce lava and ash flows, pyroclastic ejections, earthquakes, and landslides. The most disastrous volcanic eruption on record in Africa occurred at Mt Nyirangogo (DRC, January 2002), which killed 147 people and destroyed Goma, a town with over half a million inhabitants. The eruption of Mt Karthala (Comoros, April 2006) caused over 10,000 villagers to flee their homes. Mt Karthala erupts in a cycle of approximately 11 years. Previous eruptions have caused much damage to crops and pastures. Other active volcanoes that have recently erupted include Mt Cameroon (Cameroon, 1999 and 2000), Mt Nyamuragira (DRC, 1995 and 2000), Mt Fogo (Cape Verde, 2000), Mt Oldoinyo Lengai (Tanzania, 1994 and 2006), while Mt Fournaise (Réunion) remains active.

A crater lake eruption occurs when carbon dioxide dissolved in volcanic crater lakes is suddenly released into the atmosphere. This may cause mortality of people, animals, fish and insects. Over 40 lakes in Cameroon, as well as Lake Kivu in DRC, are known to contain elevated levels of carbon dioxide derived from volcanic sources. The worst recorded disaster occurred in Lake Nyos (Cameroon, 1986), where a carbon dioxide emission killed 1,876 people and numerous cattle (Lockwood and Rubin, 1989). A similar event took place in Lake Monoun in 1984, killing 37 people. Following a scientific investigation, it was recommended that the lakes be degassed. Successful tests were carried out at Lakes Monoun and Nyos in 1992 and 1995, respectively. Degassing equipment was installed at these lakes between 2001 and 2006.

Africa's preparedness for monitoring of proximal volcanic hazards and for responding to future disasters is insufficient. Systems have been installed to monitor seismic, thermal and gas emissions. These need to be complemented with satellite-based monitoring such as global navigation satellite systems and radar imagery for better mitigation strategies. There is a need to acquire remote sensing data (temperature, gases, geodetic, infrared), as well as telemetered monitoring of magnetic and electric fields, gases, temperature, etc. There is an ongoing project by the Royal Museum of Central Africa in Tervuren (Belgium) aimed at studying and monitoring African active volcanoes (SAMAAV). For this project, radar interferometry is used to study the recent evolution and assess the risks associated with four active volcanoes, *viz.* Mt Nyirangogo, Mt Cameroon, Mt Fogo, and Mt Oldoinyo Lengai. This work is done in collaboration with African volcanologists, *e.g.* University of Buea, Cameroon. However, this project only involves limited groundwork.

Landslides, Mudflows, Erosion and Siltation

Mass movements include a range of natural phenomena including landslides, mudflows, erosion and siltation, and are affected by rock and soil types, rainfall patterns, topography, and vegetation. Human factors that contribute to mass movements include overpopulation, poor land management, deforestation, and failure to enforce national physical development plans.

Landslides and mudflows cause considerable loss of life, damage to croplands, and damage of infrastructure such as highways, railways, and pipelines. Along the East African Rift, the steep topography coupled with seasonal rainfall, constitutes the main factor for generation of landslides. For example, the *El Nino* weather phenomenon in 1997-1998 caused widespread landslides and floods in parts of Kenya (Ngecu and Mathu, 1999). The national economic loss due to landslides was estimated at US$ 1 billion. In Réunion, a landslide triggered by heavy rainfall and unstable ground overran a busy coastal road in March 2006. A major transport route was destroyed, causing disruption of economic activity. Vehicles were buried, causing several deaths. Landslides are also common along the Cameroon Volcanic Line. Most are due to heavy rainfall, although earthquakes trigger some. Swarms of over 100 landslides may occur within an area of 5x5 km^2. Recent events in Cameroon include the Limbe landslides in 2001 and the Wabane landslides in 2003, where 21 and 23 people were killed, respectively. Heavy economic and infrastructural damage was also caused.

Erosion may cause degradation of arable land, with a consequent reduction of agricultural production. Siltation of rivers and dams result in shallow waters with severe implications on irrigation schemes and consequent reduction in agricultural production as has been the case in Zimbabwe. In Mauritius, deforestation has accelerated erosion. The consequent siltation has a major impact on coastal economic activities such as fishing and tourism.

An inventory of mass movements will be a valuable tool to advance research. Stabilising slopes, and enforcing land use planning in vulnerable areas can mitigate these phenomena.

Biological Hazards

Epidemics and insect infestations account for 36 per cent of all disasters in Africa (UN-ISDR, 2004). In recent decades the damaging effects of such plagues have become increasingly severe, due to the steady and continuous increase in population.

Epidemics

The ICSU-ROA Health and Human Well-being Scoping Group cover health issues. For the sake of completeness, diseases associated with environmental phenomena such as flooding and drought are mentioned here. Malaria, which is a disease carried by Anopheles mosquitoes, kills over 1.5 million people in Africa every year. An African child dies of malaria every 30 seconds. According to the World Bank reports, the mosquito-borne disease is the leading killer of African children. In 2005 there were a series of cholera outbreaks in Burkina Faso, Guinea, Guinea-Bissau, Liberia, Mali, Mauritania, Niger, Senegal and in the south of the DRC near the eastern border with Rwanda and Burundi. According to the UN's Office for the Coordination of Humanitarian Affairs, there were over 24,000 cases of cholera in West Africa in 2005. Chikungunya and dengue fever are associated with environmental phenomena such as flooding and drought. In the East African Rift, landslides may cause outbreaks of Valley Fever by releasing a fungus, found in the soil, into the air where it may be inhaled.

Pest Infestations

Pests such as locusts, grain-eating birds and African armyworm cause great agricultural losses, contributing to poverty and famine.

Astrophysical Hazards

Space Weather

Adverse space weather associated with coronal mass ejections (CME) and solar flares is a natural hazard that can cause loss of technological systems such as satellite systems, radio communication and electrical power distribution systems in the southern parts of Africa. CME's entering the Earth's upper atmosphere cause large currents (electrojets) in the ionosphere, which interact with and disturb the Earth's magnetic field. Low frequency currents may be induced in power lines, causing severe damage on power system equipment and subsequent blackouts. Many other technologies associated with infrastructure can be affected by adverse space weather including HF Radio communication, satellite communications, satellite systems, global positioning systems, and pipelines.

Africa is one of only two continents that do not yet have Regional Warning Centres (RWCs) for space weather. The Southern African Space Weather and Ionospheric Information Service (SASW) at the Institute of Maritime Technology in Cape Town provide the only known warning service in Africa. The International Space Environment Service (ISES) provides space weather predictions. ISES is a permanent service of the Federation of Astronomical and Geophysical Data Analysis Services (FAGS) under the auspices of the International Union of Radio Science (URSI) in association with the International Astronomical Union (IAU) and the International Union of Geodesy and Geophysics (IUGG). The mission of the ISES is to encourage and facilitate near-real-time international monitoring and prediction of the space environment by the rapid exchange of space environment information to assist users reduce the impact of space weather on activities of human interest.

Due to the limited impact of space weather compared to other natural hazards, there has been very little research conducted in Africa on Space Weather. The Hermanus Magnetic Observatory, situated 100 km east of Cape Town, has been approached by ISES and by the South African Department of Communications to join forces with SASW to become the regional warning centre (RWC) for space weather in Africa.

Meteorite Impacts

The African continent carries the scars of seventeen confirmed meteorite impacts, ranging in age from the 2,023 million year old Vredefort structure in South Africa, to the 10,000 year old Aorounga structure in Chad (Reimold and Gibson, 2005). The number of identified impact structures is low compared to relatively well-explored regions such as Scandanavia, and it is considered likely that more discoveries will be made in the future. While there is no instance of a meteorite impact disaster during recorded human history, the phenomenon deserves mention in this inventory, as truly catastrophic losses could result should an impact occur in a densely populated region. For example, a 50 m diameter meteorite similar to the one that created the relatively modest Tswaing crater (diameter of 1.13 km) north of Pretoria 220,000 years ago would have an explosive force equivalent to 20 to 40 million tons of TNT. Anything in the immediate target area would be instantly vaporised, and violent wind and ejecta would cause devastation over an area of 1000 km^2 or more.

Human-induced Hazards

Air and Water Pollution

Air pollution is becoming a serious environmental problem in Africa. Africa has been experiencing the world's most rapid rate of urbanization at nearly 5 per cent per annum. This, alongside tax regimes that encourage utilisation of dirty fuels, a sharp rise in the importation of old and outdated cars, and inefficient industrial plants, is increasing levels of air pollution. The high rate of urbanisation (4 to 8 per cent in some cities), expected to be sustained for the next decade, combined with low-income solutions to daily commuting, has resulted in the rapid increase in pollutants emitted by motorised vehicles. The available information suggests that the concentrations of toxic metals in many ecosystems are reaching unprecedented levels. Because of the heavy load of contaminated dusts in the air of highly-populated cities, the ambient concentrations of toxic metals are now among the highest reported anywhere. Lead (Pb) pollution from the increasing number of automobiles and cottage industries represents a major health hazard, and it is estimated that 15-30 per cent of the infants in some urban areas may already be suffering from lead (Pb) poisoning.

According to the Africa Environment Outlook (AEO) report (UNEP, 2002), the use of biomass fuel, besides degrading the environment, also impacts on the health risks of women and children who mostly do the cooking for the African families. In Tanzania, for example, children under five who die from acute respiratory infections are three times more likely to have been exposed to the burning of such fuels.

To address the issues related to air pollution, a regional network of scientists, policy-makers and non-governmental organisations, known as the Air Pollution Information Network for Africa (APINA), has been established and currently covers the southern Africa region. These activities form part of a Programme on Atmospheric Environment Issues in Developing Countries, coordinated by the Stockholm Environment Institute (SEI) and funded by the Swedish International Development Cooperation Agency (SIDA) under a project entitled "Regional Air Pollution in Developing Countries" (RAPIDC).

Water Pollution is also a serious hazard in sub-Saharan Africa. In 2000, over 300 million people did not have access to clean and safe water, and over 500 million went without adequate sanitation. Additionally, low-income urban dwellers have to pay high prices for water, sometimes up to 50 times the price paid by higher income groups. This problem has been worsened by a high rate of uncontrolled urbanisation.

Gas Flaring

Gas flaring is a serious hazard in southern Nigeria. Every day almost 2 million cubic feet of natural gas is burnt during crude oil production, more than any gas flare reported from elsewhere in the world. Not only is gas flaring a major cause of environmental pollution in the Niger River Delta, where most of Nigeria's oil is produced, but also it wastes a valuable resource. According to the World Bank report, gas flared in Nigeria is equivalent to the total annual power generation in Sub-Saharan Africa, excluding South Africa.

Mining Impacts

Mining activities may produce in a wide range of environmental hazards. Mercury is used to extract gold by thousands of small scale and artisanal miners. The mercury is initially released as gas to the atmosphere and may be inhaled by the local population. After condensation it enters the water drainage system and the food chain. No reliable estimates exist of the quantity of mercury released to the atmosphere.

Conflict-related

Africa is one of the places in the world where ongoing conflicts exacerbate other hazards. Fragile and degraded environments can fuel conflict and war and vice versa. Conflicts exacerbate the effects of natural hazards, such as famine and epidemics, by increasing the vulnerability of societies and ecosystems already under stress. In turn, the type, onset and intensity of conflicts are also influenced by natural hazards, particularly environmental hazards. Both are linked, but the relationship is complex. Therefore, these issues need to be integrated in disaster risk reduction interventions. In 1985, almost all drought-affected African countries (*e.g.*, Ethiopia, Sudan, Chad and Mozambique) were also wracked by civil wars (Timberlake, 1994). Obviously, in a conflict situation, governments allocate resources to war and put low priority to long-term environmental concerns. Landmines and unexploded ordinance affect 30 of Africa's 54 countries (Human Rights Watch, 1993 and 1994). Severely mine-affected sub-Saharan African countries include: Angola, Chad, Eritrea, Ethiopia, Somalia, Mozambique and Zimbabwe. Conflict and land degradation may cause large numbers of people to move within the borders of a country or across international borders.

Climate change

Climate change is a crosscutting issue being dealt with by the ICSU-ROA Global Change scoping group. Climate change may exacerbate many of the hazards noted above. For example, sea-level rise will cause coastal erosion and is an especially serious threat to island states. It is important to note that the prediction of future hydrometeorological events from past occurrences in no longer meaningful. The 1 in 100 year flood of the 20th century may be 1 in 50 or 1 in 20 year flood in the 21st century.

Key Challenges

Knowledge, Technology and Capacity Gaps

With a few exceptions, countries in Sub-Saharan Africa lack the capacity to conduct research on environmental hazards and disasters, or to apply the knowledge and implement technologies to mitigate environmental disasters. Compared to the developed world, there is a lack of adequate data, information, knowledge, and human resources. Furthermore, there are many other competing claims to limited resources, and the proportion of GDP devoted to scientific research lags far behind that of developed regions. Governments in Africa tend to rely on international donors rather than build indigenous research capacities. There is a need to improve training and capacity building to facilitate better use of research results in policy making. Future prospects need to be understood, and options negotiated.

Our review of recent work on natural and human-induced hazards and disasters indicates that there is a good deal of research activities in Africa. However, for most countries, there are gaps in the availability and quality of scientific data and information and often there is very little sharing of information. For example:

☆ Historical records are often inconsistent and incomplete due to difficulties in establishing and maintaining observation and data management systems. For example, several hydrometeorological stations on the major rivers of central Africa, which were perfectly operational during the colonial period and up to the 1980s, are no longer functional. Earth Resource Satellites provide excellent information relevant to studies of environmental hazards, for example, the hydrology, topography and land use of catchments, and variables such as soil moisture and snow cover. However, many countries lack funds, infrastructure, software and skills to download and interpret the data. Free data sets (*e.g.* Landsat, NOAA, Modis, Meteosat), while very useful, often has low spatial and/or temporal resolution. While Landsat data has comparatively high spatial resolution, only historical data is available, as the system has ceased to operate. More accurate high-resolution data (*e.g.* SPOT) may be very expensive.

☆ The emphasis of research into the use of geospatial information for disaster management activities should be placed more on disaster prevention and mitigation and less on emergency disaster response. Although solutions that build upon space-based systems could be used in every phase of the disaster cycle, there are clearly two types of systems that could be developed. During the crisis response phase, an "on demand" system is needed; while during the inter-crisis monitoring and warning phase an "always on" system is required. Whereas, during a crisis, high-resolution imagery is usually needed, the "always-on" system could probably be at a lower-resolution using lower cost or free imagery. That fact, together with the need to focus on vulnerability and risk analysis, supports the need for a shift from emergency response to prevention. Emphasis should be on building knowledge of the risks and monitoring, which will improve predictions of disasters and the mitigation of the impacts (UN-OOSA, 2006).

☆ There is a shortage of observation platforms that can respond rapidly to disasters *e.g.* helicopters, airplanes. It is ironic that helicopters are often made available after a disaster has happened, but are not easily procured for monitoring aimed at hazard prediction and disaster prevention.

Vulnerability and Resilience of Socio-ecological Systems

Most of the environmental problems facing Africa are difficult to solve as they are chronic, diffuse and persistent, and disproportionately affect deprived communities.

Research is needed on how to communicate warnings of impending disasters effectively, and how to disseminate knowledge to help communities improve their resilience. The values, needs and interests of different groups/stakeholders should be taken into account.

Rural communities have developed certain coping strategies. These indigenous knowledge systems should be investigated, validated and standardised, as they are often site-specific. Traditional knowledge in one region might not be applicable to another region and climate change may result in hazards beyond the knowledge bank of indigenous systems. It is essential to identify best practices for hazard/disaster reduction that can be used as a model for others. In this respect, it is necessary to improve the interaction between local, regional, national and international research teams for the successful exchange of research findings. This requires a multi-disciplinary approach. Many projects have failed in Africa because they are not based on local needs, initiatives and material resources. Therefore, research on disasters resilience and on how to tap the knowledge base of rural communities to mitigate local vulnerabilities is needed.

African is the continent where urban areas are growing the fastest and where millions of poor people are living in slums on hazardous sites such as flood plains and steep slopes. However, little attention has been given to the vulnerability of city dwellers to environmental hazards other than air pollution. Urban people suffer direct vulnerability to floods, fires, and disease epidemics, and low-income groups are particularly susceptible to drought because of their strong link to rural food supplies.

Coastal areas are increasingly being developed for shipping, security zones, recreation and tourism, fishing and agriculture, habitation and job opportunities. Small island states are particularly vulnerable to damage to natural ecosystems (coral reefs, wetlands, freshwater resources, marine resources, forests and soils) because of the small land area, limited resources, and the fragile, ocean-based economies. In coastal regions, the informal settlements of the poor are particularly vulnerable to tsunamis and storm surges along coasts. They are also likely to bear the brunt of rising sea levels. Detailed information on present and future population exposure to coastal hazards exposure is essential for mitigation policies and technologies.

Vulnerability and Resilience of Technological Systems

All countries, including those in Sub-Saharan Africa, are dependant on their power transmission and information technology infrastructure, and the level of dependence is likely to increase as African countries seek to bridge the "digital divide". Many natural hazards, such as floods, earthquakes, and space weather, can damage these technological systems causing widespread chaos and economic loss. The vulnerability of these systems must be assessed, and steps taken to improve their resilience.

Effective Transfer of Information to Policy Makers

There is a need to establish dialogue between scientists, policy and decision makers. As environmental degradation is not only a technical/scientific problem, any discussion of environmental degradation should involve policy and decision makers. Research is needed on how to translate research results into policies that minimise the human and economic cost of hazards; for example, in land use planning and environmental issues. In this respect, ICSU ROA can play an important role in promoting and linking scientific research and capacity building in Africa to policy and decision makers and society.

Integrated Modelling of Multiple Disasters

A research and implementation plan on natural and human-induced hazards requires an integrated, multi-hazard approach (*e.g.* environment degradation, conflicts, health hazards) that addresses vulnerability and risk assessment as an integral component of disaster management. Implementation strategy should also provide the scientific scope for the reduction of the risks and consequences of natural and human-induced environmental hazards. Integrated environmental and socio-economic modelling and scenario building are needed to identify the scale and direction of the necessary mitigating and recovery strategies. In this respect Earth observation satellites and geoinformation technology are valuable tools for hazards managers and respondents. By using multiple modes of observation, researchers can create methods for integrating information from various sources.

Early Warning and Preparedness

Early warning is the provision of timely and effective information through identified institutions, to allow individuals exposed to a hazard to take action to avoid or reduce their risk and prepare for effective response. There is an urgent need to transmit scientific knowledge on hazards to support early warning and preparedness. The challenge is how to provide relevant education at different levels (communities, schools, tertiary institutions) to facilitate mitigation of hazards. A gender perspective is also essential in disaster risk management policies, plans and decision making processes, including those related to risk assessment, education and training. For example, to address increasing disasters from natural hazards, the UN International Strategy for Disaster Reduction (ISDR) report on Global Survey of Early Warning Systems (ISDR, 2006) proposed a People-Centred Early Warning Systems (P-CEWS) that is relevant for addressing disaster management in Africa. The aim of the P-CEWS (Figure 1.4) is to "empower individuals and communities facing hazards to act in sufficient time and in appropriate manner to reduce personal injury, loss of life, damage of property, Environment and loss of livelihoods."

It is necessary to improve the capability to identify indicators of physical, social, and environmental vulnerabilities throughout Africa and to select and implement realistic solutions to reduce them to acceptable levels. It is also necessary to develop a vulnerability index using hazard maps. This can be used by policy makers to make informed decisions and by donors to provide the required assistance.

Environmental Change

There is consensus among scientists that climate change is a growing threat. However, questions on how climate change will directly impact risk patterns remain largely unanswered because current climate models are unable to predict specific

---

An Effective People Centered Early Warning Systems is formed by 4 inter-related elements

Risk Knowledge	Monitoring & Warning Services
Systematically Collect Data and Undertake Risk Assessment	Develop Hazard Monitoring and Early Warning Services
Dissemination and Communication Communicate Risk Information and Early Warnings	Response Capability Build National and Community Response Capabilities

Weakness/failure in any part of these elements may result in complete failure of the whole systems

---

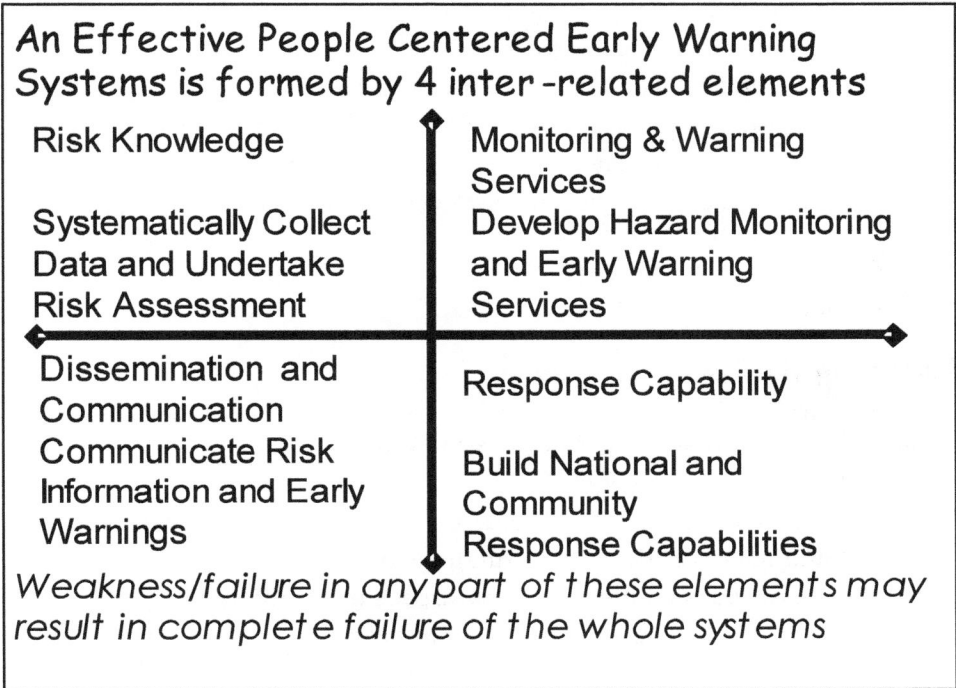

Figure 1.4: People-Centred Early Warning Systems (ISDR, 2006).

alterations in weather patterns, storm severity, or habitat degradation. However, there is emerging work on drought and heat waves. Mitigating the unpredictable outcomes of climate change presents a difficult challenge for society at every scale. There is a need to determine how to integrate adaptation to, and preparedness for risks of hazards such as floods, droughts, tsunamies etc, resulting from climate change.

Environmental Degradation

Long-term environmental problems can fuel conflicts and civil wars. Conflicts and wars can also contribute to environmental degradation. But the interconnections are complex. Research is required on the link between environmental degradation and conflicts.

Proposed Research Activities

Four flagship projects are proposed. All projects are multi-disciplinary and have been defined in terms of geographic regions that transcend national borders.

Vulnerability Science

This overarching project addresses generic issues relevant to the mitigation of environmental hazards and disasters. Issues to be addressed include:

☆ The development of a multi-hazard database,

☆ Integrated modelling of multiple disasters,

☆ Early warning and preparedness,

☆ Assess and develop methodologies *e.g.* decision-support tools,

☆ Establishing a index describing vulnerability to compounded disasters,

☆ Compilation of vulnerability maps,

☆ Risk analysis techniques and disaster management strategies appropriate to different risk profiles,

☆ Building of institutional capacity, and

☆ Outreach.

Mitigating the Risk of Flooding

Floods are a major hazard in Africa. Watersheds would be used to identify the countries involved in a particular study. Issues to be addressed include:

☆ Vulnerability to floods,

☆ Scenarios of floods related hazards in future *e.g.* under climate change,

☆ Contribution of information to the multi-hazard database,

☆ Assessment of multi-dimensional aspects (*e.g.* early warning),

☆ Pilot study of a river catchment system that would involve several countries and include several settings (*e.g.* both rural and densely populated areas),

☆ Water politics,

☆ Capacity and institutional building, and

☆ Outreach

Mitigating the Risk of Drought

Drought has serious implications for livelihoods and environment, and is one of the causes of famine. Issues to be addressed include:

☆ Contribution of information to the multi-hazard database,

☆ Assess multi-dimensional aspects, such as the future vulnerability to drought, spatial and temporal trends of drought in the continent; drought and land degradation; and fire and drought,

☆ Pilot study of a region vulnerable to drought (*e.g.* SADC, East, West and Central Africa, the Sahel region) that includes several settings (*e.g.* both rural and densely populated areas),

☆ Capacity building, and

☆ Outreach.

Geohazards

It was proposed that ICSU-ROA to mobilise the African scientific community to submit an expression of interest and formulate a research and outreach proposal under the hazards theme of IYPE. The proposed scope of work includes:

☆ Crater lakes,

☆ Mineral resource exploitation issues (mercury pollution, oil spills and rockbursts),

☆ Rift Valley earthquakes and landslides, and

☆ Book on geohazards.

A proposal was formulated that addressed earthquake and volcanic hazards, but at the time of writing (2010) had not been successful in raising support. ICSU-ROA is actively supporting the Sub-Saharan Regional Programme of the Global Earthquake Model (GEM, 2010) to address earthquake risks.

Conclusions

The mobilisation of Africa's intellectual resources will undoubtedly be the critical factor in ensuring implementation of the ICSU ROA science plans for Africa. ICSU ROA can serve as a bridge between African institutions and the international scientific community. In this respect, the need for African scientists to work with local communities in evaluating risks and finding ways to respond to risks cannot be overemphasised. To this end, the following recommendations were made.

Science Priorities

1. A database and set of analysis tools for the prediction and mitigation of environmental hazards, risk reduction and disaster management should be developed.

2. ICSU ROA should ensure the establishment of mechanisms to monitor progress of activities in Africa with regard to hazard preparedness and mitigation.

3. The effect of climate change on various hazards (floods, fire, etc.) should be addressed through research targeted at hazard early warning and vulnerability/resilience of socio-economic systems. In pursuance of this, ICSU ROA should leverage the gains of past and ongoing research activities of several development partners (World Bank, CIDA, etc) in capacity building and needs assessment.

Partnerships

1. For its successful implementation, the science plan needs the support of African scientific institutions as well as regional and international partners. ICSU ROA should work to strengthen regional national, and international institutional frameworks to facilitate disaster risk-related information management and sharing. Coordination of different hazards and disasters research initiatives at regional levels should be facilitated.

2. Information regarding research activities (*i.e.*, scientific data, reports and publications) should be disseminated through existing networks such as NEPAD, SADC, ECOWAS, IGAD, SEMAC, IOC, the AU and other pan-African structures.

Capacity Building and Outreach

1. Centres of Excellence should be supported in efforts to offer specialised courses to practitioners and to involve postgraduate students in research projects.

2. Education and awareness-raising campaigns should be directed, as far as possible, at all stakeholders at all levels and using all structures, to ensure understanding of warnings of forthcoming hazards and disasters.

3. It is also necessary to introduce key research findings into school and tertiary curricula by developing teaching aids, for example, DVDs, CDs, and posters. On-line computer-aided interactive learning modules should be developed. For example, case histories with real data and tutorial exercises (an on-line module is being developed by Universities in Mauritius, Malta and South Pacific dealing with vulnerability of islands to natural disasters). The African Virtual University (AVU) in Nairobi is developing teaching materials. UNISA (a South African distance learning institution) offers a module in Disaster Management. The University of Botswana has established policy on "virtual centres" to link scientists working on specific themes, including environmental hazards and disasters.

Acknowledgements

We thank the staff at the Regional Office for Africa of the International Council for Science (ICSU) for their support. They are Prof. Sospeter Muhongo (Director), Dr. Achuo Enow, Mrs. Masela Pillay and Mrs. Kathy Potgieter. We also thank Dr. Khotse Mokhele (President of the South African National Research Foundation) for his direction and support. Dr. Mokhele is Vice-President of ICSU. He was part of the team responsible for drafting the ICSU Strategic Plan for 2006-2011, and is Chairman of the Committee for Scientific Planning and Review.

References

[1]   Anyamba, A. and Tucker C. J. (2005) Analysis of Sahelian vegetation dynamics using NOAA-AVHRR NDVI data from 1981–2003. Journal of Arid Environments, 63: 596–614.

[2]   Burke, E. J., Brown, S. J. and Christidis, N. (2006). Modelling the recent evolution of global drought and projections for the 21st century with the Hadley Centre climate model. Journal of Hydrometeorology, 7: 1113-125.

[3]   Díaz J., Jordán, A,. García, R., López C., Alberdi J., Hernández, E. and Otero A. (2004) Heat waves in Madrid 1986-1997: effects on the health of the elderly. Journal International Archives, 75: 163-170.

[4]   Dilley M., Chen, S., Deichmann, U., Lerner-Lam, A. L. and Arnold, A. (2005) Natural Disaster Hotspots: A Global Risk Analysis, World Bank.

[5]   Easterling, D. R., Evans, J. L., Ya, P., Groisman, P Ya, Karl, T. R., Kunkel, K. E., and Ambenje, P. (2000), Observed Variability and Trends in Extreme Climate Events: A Brief Review, Bulletin of the American Meteorological Society, 81: 417-425.

[6]   Economist (2006) The Economist Intelligence Unit, http://www.viewswire.com, 11 May 2006.

[7]   Foley J. A., Coe M. T. and Wang G. (2003) Regime shifts in the Sahara and Sahel: Interactions between Ecological and Climatic Systems in Northern Africa. Ecosystem, 6: 524-539.

[8]   Freiburg University (2006) http://www.fire.unifreiburg.de/current/archive/archive.htm.

[9]   GEM (2010) Global Earthquake Model http://www.globalquakemodel.org/.

[10]   Goldammer, J. G. and De Ronde, C. (eds) (2004) Wildland Fire Management Handbook for Sub-Sahara Africa. Global Fire Monitoring Centre (GFMC).

[11]   Herrman, S. F., Anyamba, A. and Tucker, C. J. (2005) Recent trends in vegetation dynamics in the African Sahel and their relationship to climate, Global Environmental Change, 15: 394-404.

[12]   Holloway, A. (1999) Disaster Awareness and Public Education in Africa, Natural Hazards Observer, 23(6).

[13]   Human Rights Watch (1993) Landmines in Angola. Human Rights Watch Arms Project and Human Rights Watch/Africa, New York.

[14]   Human Rights Watch (1994) Landmines in Mozambique. Human Rights Watch Arms Project and Human Rights Watch/Africa, New York.

[15]   ICSU (2005) Natural and human-induced environmental disasters. Report from the ICSU Global Scoping Group.

[16]   IPCC (2001) Africa Chapter 10, Climate Change. Impacts, adaptation and vulnerability. Intergovernmental Panel on Climate Change. Cambridge University Press.

[17]   ISDR (2006) Global Survey of Early Warning Systems. A report prepared at the request of the secretary-general of the UN. Pre-print version released at the Third International Conference on Early Warning, Bonn, 27-29 March, 2006.

[18]   Karl T., Linda, R, and Mearns, O. (2000) Climate Extremes: Observations, Modelling, and Impacts. Science, 289: 2068-2074.

[19]   Lockwood, J. P. and Rubin M. (1989) Origin and age of the Lake Nyos maar, Cameroon: Journal of Volcanology and Geothermal Research, 39: 117-124.

[20]   Meehl, G. A. and Tebaldi, C. (2004) More intense, more frequent, and longer lasting heat waves in the 21st century. Science, 305: 994-997.

[21]   Mulugeta, G., Ayonghe, S., Daby, D., Dube, O. P., Gudyanga, F., Lucio, F. and Durrheim, R. J. (2007). Natural and Human-induced Hazards and Disasters in Sub-Saharan Africa, ICSU Regional Office for Africa, 30 pp.

[22]   NASA (2006) http://rapidfire.sci.gsfc.nasa.gov.

[23]   NEIC (2006) http://neic.usgs.gov/neis/general/seismicity.

[24]   Ngecu, W. M. and Mathu, E. M. (1999) The *El Nino* triggered landslides and their socio-economic impacts on Kenya. Episodes, 22: 284-288.

[25] Nicholson, S. E. (2001) Climate and environmental change in Africa during the last two centuries. Climate Research. 4: 123-144.

[26] Nicholson, S. E. (2005) On the question of the "recovery" of the rains in the West African Sahel. Journal of Arid Environments, 63: 615-641.

[27] OFDA-CRED (2002) Centre for Research on the Epidemiology of Disasters (CRED) International Disaster Database 2002: http://www.em_dat.net/disasters/profiles.php.

[28] Pyne, S. J., Goldammer, J., De Ronde, C., Geldenhuys, C. J., Bond, W. J. and Trollope, S. W. (2004) In Goldammer, J. G. and de Ronde, C. (eds), Wildland Fire Management handbook for Sub-Sahara Africa. Global Fire Monitoring Centre (GFMC).

[29] Reimold, W. U. and Gibson, R. L. (2005) Meteorite Impact! Chris van Rensburg Publications, Johannesburg.

[30] Robinson, P. J. (2001) On the Definition of a Heat Wave. Journal of Applied Meteorology, 40: 762–775.

[31] Scholes, R. J. and Biggs, R. (eds) (2004) Southern African Millennium Ecosystem Assessment. Ecosystem Services in Southern Africa. A regional assessment. CSIR.

[32] Semazzi, F. H. M. and Song, Y. (2001). A GCM study of climate change induced by deforestation in Africa. Climate Research, 17, 169-182.

[33] Swiss Agency for Development and Cooperation (2006) Africa's Ecological Footprint: Human Well-Being and Biological Capital, http://www.footprintnetwork.org/Africa.

[34] Timberlake, L. (1994) Africa in crisis: The causes, the cures of environmental bankruptcy. East African Educational Publishers, 203 pp.

[35] UNEP (2002) Africa Environmental Outlook: Past, present and future perspectives, www.grida.no/aeo/index.htm.

[36] UNESCO-IOC (2005) Assessment of capacity building requirements for an effective and durable tsunami warning and mitigation system in the Indian Ocean: consolidated report for 16 countries affected by the 26 December 2004 tsunami. UNESCO-IOC Information Document No. 1219, UNESCO, Paris.

[37] UNICEF (2006). http://www.unicef.org.uk/emergency/.

[38] UN-ISDR (2004) Towards sustainable development in Africa. Report on the status of disaster risk management and disaster risk assessment in Africa. UN/ISDR, Africa Development Bank, African Union, New Partnership for Africa's Development, 57 pp.

[39] UN-OOSA (2006) http://www.unoosa.org/pdf/reports/ac105/AC105_808E.pdf.

[40] Webb P., Braun, J. and Teklu, T. (1991) Drought and famine in Ethiopia and Sudan: An ongoing tragedy. Natural Hazards, 4: 85-86.

Chapter 2

Early Warning Systems for Natural Disasters in The Republic of Mauritius

S.N. Sok Appadu

Mauritius Meteorological Services,
Vacoas, Mauritius
E-mail: *meteo@intnet.mu, luxmi80@hotmail.com*

ABSTRACT

After giving a brief introduction to the Mauritian economy, the efforts being done in Mauritius to predict and warn the general public about the natural disasters are described. The Early Warning Systems (EWS), in use in the Republic of Mauritius, have been undergoing continuous changes to take into account new extreme events, additional information being made available, new technologies and of the evolving needs of socio-economic activities and developments.

Points out that the Mauritius Meteorological Services (MMS) has been at the forefront of providing EWS to the nation through the operation of a 24-hour, seven-day a week and year round system for monitoring, detecting and forecasting natural hazards. Preventive measures and preparation of an action plan in case of occurrence of a natural disaster is also explained.

It can be concluded that with the assistance of the government, the National Meteorological Services is bringing together stakeholders from a wide range of users of early warnings, which are people- centred and are based on the needs, priorities, capacities and cultures of those at risk. People are being encouraged to be partners in the system, not to be controlled by it.

Introduction

The Republic of Mauritius consists of a main island, Mauritius, and a group of small islands scattered in the tropical waters of the South West Indian Ocean. These islands are Rodrigues, the Cargados Carajos (St Brandon), Agalega, Tromelin and

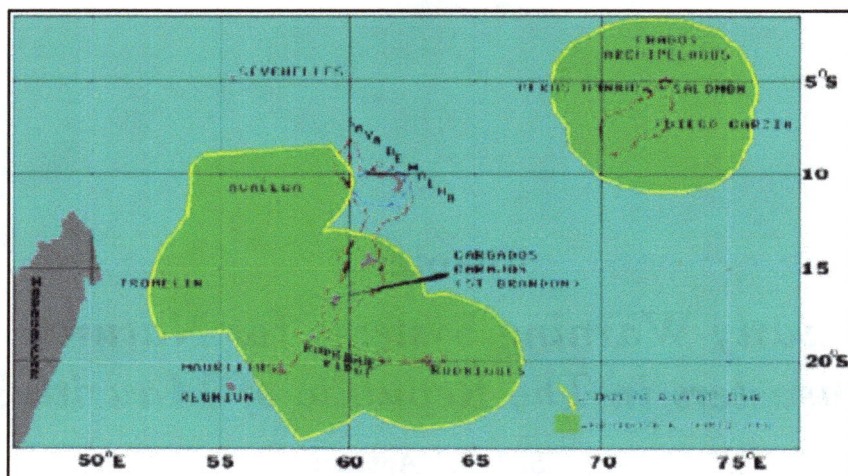

Figure 2.1: The Republic of Mauritius's Economic Zone.

the Chagos Archipelago (Diego Garcia) (Figure 2.1). The total land area of the Republic is 2040 km^2 and is surrounded by coral reefs. The main island is situated about 2000 km off the East Coast of Africa.

Mauritius and Rodrigues are the results of four major volcanic activity periods between 7.8 M and 25,000 years ago. The other islands are mostly sand banks and do not rise more than 5 metres above sea level. Mauritius and Rodrigues lie near the edge of the southern tropical belt and are free from the influence of large land masses or continents. They enjoy a mild maritime climate with summer extending from November to April and winter from May to October. May and October are transition months during which the weather is variable.

Agalega, St Brandon, Tromelin and Diego Garcia lie within the tropical belt of the South Indian Ocean. They are directly along the route of tropical cyclones and the inter-tropical convergence zone. They are not highly inhabited. St Brandon suffers from the direct effects of tropical cyclones, so much that the configuration of the island rarely remains the same after a cyclone event.

The Republic is densely populated with more than 550 persons/km. Urban population represents about half of the total population. Life expectancy, 70 years, is among the highest in Africa with male reaching above 66 years and females 74 years. The population growth is stabilizing at around 1 per cent . Infant mortality rate has continued to decline and is now standing at less than 20 per thousand.

Mauritius has established a comprehensive social welfare system with a wide range of assistance schemes such as old age pensions, basic widow's pensions, basic invalid pension and other non-contributory social benefits to cater for the needy. There are programmes, for poverty alleviation to ensure that the poor are not excluded from the main stream of socio-economic development. The entire population has access to free medical services that has resulted in a very significant improvement in the health status of the population. The number of households having access to electricity and potable water has crossed the 90 per cent mark.

The education system is a four-tiered one, starting with the pre primary, primary, secondary and tertiary levels. The education system is free throughout and is obligatory for all children between the ages of 4 to 11. Adult, literacy rate is about 90 per cent . The Ministry of Education and Human Resource is laying strong emphasis to further promote science subjects at different levels in the national curriculum. Professional and Vocational Training are also offered by other institutions.

The prospect for sustained development of the new independent Mauritius, 1968, was rated quite bleakly by eminent world economists. However Mauritius went on to record a robust average annual growth rate of about 5 per cent over the last 30 years due to relentless commitment by successive Governments to a consistent programme of economic reform and liberalization. The economy has undergone several distinct phases and has diversified from a mono coop economy, highly dependent on the export of sugar, into exports, manufacturing, tourism, and more recently services.

The manufacturing sector has been the major driving force of the economy. The rapid growth in this sector was concentrated in the Export Processing Zone. This sector is dominated by textiles and garments and provides bulk employment to the non professional population. The Sugar Industry has been the backbone of the Mauritian economy since the dawn of its economic history. It is still an important economic player, although in the diversification process, it has been overtaken by the other sectors, especially tourism and services.

The Tourism Industry has established itself as, another of the main cylinders of growth. The main source markets for the tourist sector continue to be the Europeans, followed by South Africans. A breakthrough has been noted with more and more tourists, coming from India, China, South East Asia and Australia. Mauritius will continue to promote itself, as a quality destination catering to the long haul high spending end of the market.

The quaternary sector, comprising the new high-tech international financial services such as off-shore banking, freeport, fund management, stock exchange, insurance, information and communication technology, sea-food hub, has emerged to be vital contributors to the economy and is now considered as a major pillar. Its contribution to GDP is higher than the agricultural sector and at par with the EPZ and Tourism Sectors. The Off-shore Business Centre has given a whole new shape to the structure of the financial system and has broadened the scope of financial activities. Mauritius has acquired the stature of a trustworthy, stable and reputable offshore jurisdiction. The Mauritius Freeport has also witnessed an impressive growth performance. The major activities of the Freeport are, inter alia, warehousing and storage, labelling, packing, breaking bulk now other processing activities, including the seafood hub. Mauritius is now well poised to be Africa's major regional Freeport centre and the ideal trans-shipment port to interface the growth centres of Asia to the burgeoning economies of the African Continent.

Early Warning Systems in Mauritius

The Republic of Mauritius has a long history and tradition in the provision of early warnings to its population, during adverse weather events. With the advent of railway system in the Republic in the 1860s and the newly established Royal Alfred Observatory to the North, a boost was given to information sharing between the Observatory and the population. Communications through telegraphy between the Observatory and the main railway station, in Port Louis, was established, then to all railway stations scattered over the island.

The Early Warning System, in use in the Republic of Mauritius, has been undergoing continuous changes to take into account new extreme events, additional information being made available, new technologies and of the evolving needs of socio-economic activities and developments. As from 26 December, 2004, with the advent the ocean-wide tsunami in the Indian Ocean, Early Warning System has been also extended to events other than natural disasters generated by meteorological and hydrological conditions. The EWS is not static, it is dynamic, since it has to be continuously updated and amended to accommodate new arising situation and it has to cater for the best interests of the population.

Early Warning Systems have the potential to prevent natural hazards from becoming disasters. Resulting activities may considerably reduce loss of life and socio-economic damages. Disaster preparedness and management, urgently need the establishment of a well functioning EWS, which will be in a position to provide timely, accurate and precise information to population at risk.

Mauritius Meteorological Services, MMS, has been at the forefront of providing EWS to the nation through the operation of a 24-hour, seven-day a week and year round system for monitoring, detecting and forecasting natural hazards. Warnings are systematically issued during extreme weather events such as tropical cyclones, torrential rains, flash flood, lightning, drought and associated events such as landslides, landsteps, pest invasion and impacts on health.

Emergency planning and response require the collaboration and coordination from international, regional and national levels. The Central Cyclone and Other Natural Disaster Committee, under the aegis of the Prime Minister's Office, prepare National Action Plan with the community as the focus of all actions. Guidance has been prepared and given to individual and families, as well as "Les forcecs vives" to prepare personal action plans, lists of emergency supplies to have on hand and a clear understanding of coordination for their own safety. A wide range of products and services are being given to the public to understand EWS in force.

Existing Warning Systems

Tropical Cyclone

Tropical cyclones are low pressure systems which in the southern hemisphere have well-defined clockwise wind circulation spiralling towards the centre with great intensity. The strongest winds as well as the heavier rains occur in the region close to the centre. The centre of the tropical is commonly known as the eye. It is an area characterised by light winds and often clear skies.

CLASS I	**is issued 36 - 48 hours before Mauritius or Rodrigues is likely to be affected by gusts reaching 120 km/h.**
CLASS II	**is issued so as to allow, as far as practicable, 12 hours of daylight before the occurrence of gusts of 120 km/h.**
CLASS III	**is issued to allow, as far as practicable, 6 hours of daylight before the advent of 120 km/h gusts.**
CLASS IV	**is issued when gusts of 120 km/h have been recorded in some places and are expected to continue.**
TERMINATION	**There is no longer any risk of gusts exceeding 120 km/h.**

Figure 2.2: Warning Issued Related to Speed of Gust.

The eye of the tropical cyclone is surrounded by a dense ring of cloud known as the eye wall which may have a width of 35 to 100 km. The maximum speed of gusts may exceed 200 km/h and rainfall up to 50 cm per day may be expected. Spiral bands which contain rain and strong winds may be located as far as 500 km or more from the centre.

Tropical cyclones derive their energy from the warm tropical ocean. They generally form on or near the Inter-tropical-Convergence Zone (ITCZ), in the belt between 05° latitude over the open sea, in an area where the sea surface temperature exceeds 26.50 C. The cyclone season extends from 01 November to 15 May in Mauritius and Rodrigues.

In facing the potential hazard of the cyclone, the Government of Mauritius has exposed precautionary warning to the public. The public stands in front of the incoming cyclone warn to take following actions:

Before the Cyclone

1. Ensuring that house is in good condition and can withstand cyclone gusts;
2. Trimming tree branches likely not to cause damages to house, telephone and electricity lines;
3. Clearing property of loose materials that can cause injury and damage during extreme winds;
4. Identifying secure places for boat;
5. Be acquainted with the nearest cyclone refugee centres;
6. Prepare an emergency kit consisting of:
   (*a*) Portable AM/FM radio and fresh batteries.
   (*b*) Torch, lamps, candles, matches, etc.
   (*c*) Water containers.
   (*d*) Canned food, can opener, stove with sufficient gas.
   (*e*) Rice, flour, biscuits, cheese, etc.
   (*f*) First aid kit and essential medicines.
   (*g*) Clothes secured in plastic bags.
   (*h*) Tool kit for emergency repairs (hammer, nail, rope, etc.).

During a Cyclone Warning Class I

1. Making sure that emergency kit is ready;
2. Monitoring cyclone bulletins on Radio/TV;
3. Preparing to secure windows and doors with shutters or shields.

During a Cyclone Warning Class II

1. Verifying that emergency kit contains all essential items;
2. Storing sufficient amount of drinking water;
3. Continuing to monitor cyclone bulletins on Radio/ TV.

Upon the Issue of a Cyclone Warning Class III

1. Complete all preparatory measures:
   (*a*) Fix shutters;
   (*b*) Secure doors and windows;
   (*c*) Store loose articles.
2. Avoiding areas prone to storm surges and flooding;
3. Sheltering domestic animals;
4. Securing vehicles;

5. Those, in insecure dwellings, move as early as possible to a cyclone refugee centre with emergency kit;

6. Avoiding going outside;

7. Monitoring closely cyclone bulletins on radio/ TV.

During a Cyclone Warning Class IV (*Gusts Of 120 km/H Or More Are Occurring*)

1. Stay inside. Seek shelter in the safest part of the house.

2. Disconnect all electrical appliances.

3. Listen attentively to cyclone bulletins and advice on the Radio/ TV.

4. If the house starts to suffer important damages, protect yourself with mattress, rugs or blankets.

Passage of the 'Eye' of a Cyclone

Beware of the passage of the 'EYE'. Do not assume that cyclonic conditions are over. The calm period is always followed by violent winds from the opposite direction.

After the Cyclone

1. Do not leave shelter until the all-clear signals have been given by relevant Authorities;

2. Beware of fallen power lines, damaged buildings and trees and flooded water courses;

3. Do not consume fallen fruits;

4. Boil water for drinking purposes;

5. Clean yard and drain out stagnant water to prevent proliferation of mosquitoes/ diseases.

Torrential Rains

The Mauritius Meteorological Services (MMS) is in a position to forecast torrential rain well in advance to cater for precautionary measures. Whenever the climatic condition over Mauritius or Rodriguez produces 100 mm of widespread rain in less than 12 hours and is likely to persist for several hours, the MMS will issue warnings at regular intervals. If such a condition prevails at the beginning of a school day, schools will remain closed. However if it is observed during school hours, the public will be informed accordingly and classes will be dismissed.

Tsunami

The tragic event of 26th December, 2004 attracted the world's attention to the vulnerability of our shoreline to natural catastrophe. The seismic event which generated devastating tsunami waves took the countries bordering the Indian Ocean by surprise and in a real state of unpreparedness.

Tsunamis, which can occur at any time and cannot be predicted, are a global, high-fatality, low-frequency hazard that can strike in minutes, and cause damage for hours and over an entire ocean basin. An effective tsunami early warning system is achieved when all persons in vulnerable coastal communities are prepared and respond appropriately, and in a timely manner, upon recognition that a potentially destructive tsunami is coming.

Institutional Framework

The Major lesson learnt is that this disaster left such a sequel of death and destruction because of a fundamental institutional failure. Although the Science and technology exist to warn, prepare for and mitigate of such a catastrophic disaster, none of the necessary programmes and institutions needed were in place in the Indian Ocean region. In addition to the need to adopt preventive measures against ocean-generated Tsunami disasters, the Republic of Mauritius whose economy and food security depend largely on the ocean and coastal areas through the tourism industry must also be able to address local problems in order to achieve a disaster resilient standard.

To stimulate the development of our scientific and technological capabilities all the Key scientific organizations under the supervision of the PMO engaged themselves to focus on the preparation of a national action plan in order to protect the local community from the hazards and improve coastal governance (Figure 2.3).

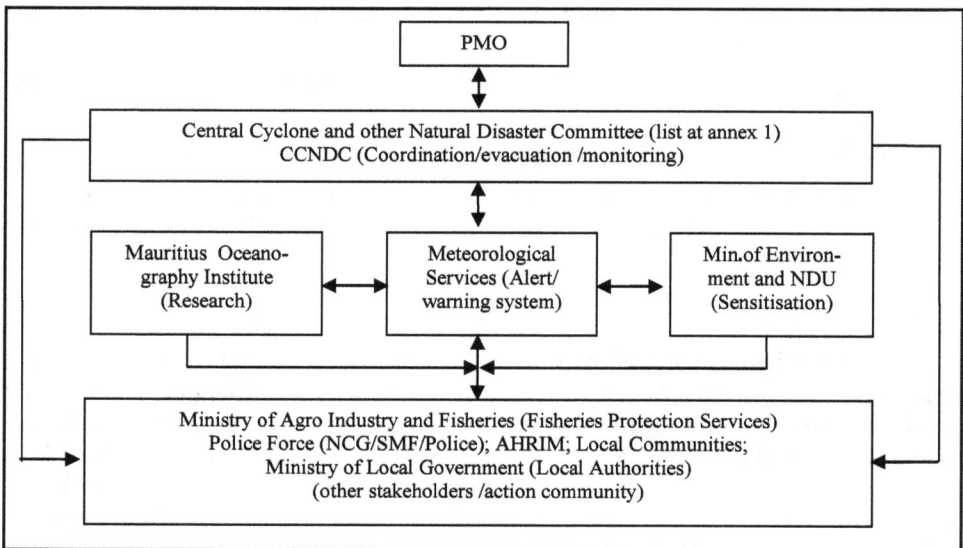

Figure 2.3: PMO Structure.

Component of the Action Plan

Coordinating Activity

Coordination and monitoring should be under the responsibility of the Prime Ministers Office (PMO) (Central Cyclone and other Natural Disasters Committee). The CCNDC is comprised of several stakeholders involved in management of the coastal zone (a list of the members involved is at Annex 1). Upon receipt of advisories on tsunami generation, the Meteorological Services will accordingly inform the PMO, which in turn will contact all the relevant stakeholders and initiate appropriate actions. Post-tsunami assessment will be carried out by the CCNDC.

Warning System

The Tsunami Warning System in Mauritius issued by the Mauritius Meteorological Services is analogous to those issued by Hawaii State Civil Defence. It consists of bulletins depending upon the severity of the Tsunami. Tsunami Warning System will consist of data reception, analysis and forecasting, warning preparation, communication of warning and community response.

The following categories of bulletin are analysed and prepared at the Tsunami Warning Centre.

1. *Tsunami Information Bulletin*: Strong earthquake has occurred in the Indian Ocean but does not constitute an Indian Ocean wide distant Tsunami threat. No action undertaken;

2. *Tsunami Advisory Bulletin*: A strong earthquake has occurred in the Indian Ocean and the generation of a destructive of an Indian Ocean Tsunami is being investigated by World Centres, PTWC and JMA. Event is closely monitored internally for possible upgrade to a watch status;

3. *Tsunami Watch Bulletin*: A major earthquake has occurred in the Indian Ocean and World Centres is evaluating the situation. Tsunami Watch contacts at national level notified and requested to be ready for appropriate action;

4. *Tsunami Warning Bulletin*: Major earthquake in the Indian Ocean and destructive Tsunami detected. Warnings disseminated to the general public through media. Tsunamis watch contact to start evacuation of vulnerable areas. Bulletins are upgraded in light of observations and updated advisories of PTWC and JMA;

5. *Tsunami Cancellation Bulletin*: No threat of destructive Tsunami wave or wave action has ceased. Travel time of Tsunamis from the 2 major sources in the Indian Ocean namely, the Indonesia and Makran region is of the order of 7 to 9 hours. Advisories received from PTWC or JMA allows enough of lead time to warn the population.

Risk Assessment

The Republic of Mauritius has recognized the importance of assessing its vulnerability in the event of a tsunami. The expert IOC/UNESCO mission to assess requirements and capacity in Mauritius (August 2005), set the elaboration of an inundation risk map as a priority. Such a tool will assist the issuance of a tsunami warning for the Republic of Mauritius.

Much effort is being deployed in numerical simulations of tsunami propagations. The Center for Tsunami Inundation Mapping Efforts (TIME) was created to assist the Pacific States in the development and maintenance of maps which identify areas of potential tsunami flooding. In the same effort the International Tsunami Information Centre (ITIC) together with other collaborators organized the first tsunami numerical modelling course. The Republic of Mauritius has benefited from such training and has now produced a first version of a tsunami preparedness map by the MOI, with the collaboration with of Key stakeholders.

Mitigating Measures and Public Awareness and Education

Although a tsunami cannot be prevented, its impact can be mitigated through community and emergency preparedness, timely warnings, effective response, and public education. There should be continuous and sustained awareness activities to sustain preparedness. Education is fundamental to building an informed citizenry and to ensure that the next generation of people is equally prepared. The Ministry of Environment and NDU with the collaboration of the Meteorological Services have already initiated the awareness campaign.

For longer-term sustainability of the tsunami education and public outreach (EPO) campaign, tsunami awareness and education is being included in the environmental campaigns currently being undertaken by the Ministry of Environment and NDU. The following activities are being planned in the context of the tsunami awareness campaign and education programme:

1. maintaining an up to date awareness programme through pamphlets and printing the education materials in Creole, French and English languages;
2. outreaching activities through talks, seminars, symposiums , electronic and press media;
3. carrying out simulation exercises/drills to test population response and evacuation strategy;
4. preparation of guiding materials with respect to construction of building along the coast and conservation of marine biodiversity.

Network and Telecommunication Systems

Network of Observing System

The current seismic and sea-level network is very poor in the southwest Indian Ocean. Apart from the two recently upgraded sea level stations located at Port Louis and Rodrigues and a wave-rider buoy off the south east coast of Mauritius there are no other real-time equipment to provide data for estimating the incoming Tsunami wave.

As a follow-up of the Hyderabad meeting, December 2005, IOC has decided to support/provide the Meteorological Services with a seismic monitoring system. This seismic monitoring system will form part in the core seismic network in the Indian Ocean. The system could be used for an evaluation of a local or a distant Tsunami. Coastal automatic weather observing System is already in place around the island of Mauritius and Rodrigues to provide real-time observation during extreme events.

Telecommunication Facilities

Global Telecommunication systems: The following Global Telecommunication system is present in Mauritius.

☆ Transmet message switching systems;

☆ Synergy visualization work systems;

☆ Global Telecommunication system link, 9.6 kbps, Reunion;

☆ GTS satellite based component, Retim – Africa receiving system;

☆ Eumetcast Receiving system.

It has been observed that advisory messages from PTWC and JMA are received instantaneously at the tsunami warning centre (MMS).

National Telecommunication System

All important key stakeholders including media, both print and electronic, are linked to MMS through phone, mobile, fax and email. The police information room can be used to disseminate alert messages to the population in remote areas. Communication facilities such as WMO GTS and AFTN can be used to provide alert messages to countries west of Mauritius. HF Radio Communication can be used to warn outlying islands namely St Brandon, Agalega and Rodrigues.

Networking among Key Players

The central cyclone and other natural disaster committee including Tsunami (CCNDC) is the national platform to encourage networking among key players. Regular meetings are being held to discuss all matters related to Tsunami including the Action Plan to develop and maintain a Tsunami Warning System for the Republic of Mauritius.

Regular forum are being organised through mechanism established by the national Meteorological Services, NODC, NOSF, to keep abreast of development in the Science of Tsunami, mitigating its impacts and sensitizing the public

Workshop/Symposium

Officers attending regional or International workshop/symposium are being requested to share knowledge and information among key players. Participants would have to submit mission reports to the Coordination Body.

Landslide

The following criteria for the issue of warnings to the inhabitants of Land Slide areas will henceforth apply:

☆ Geomorphology;

☆ Identification of landslide areas;

☆ Rainfall recording; and

☆ Ground displacement.

The warning/evacuation system shall consist of five stages as follows:

☆ Stage 1: Preparatory Stage

☆ Stage 2: Warning Stage

☆ Stage 3: Evacuation Stage

☆ Stage 4: Emergency Stage

☆ Stage 5: Termination

Stage 1: Preparatory Stage

Stage 1 is reached when rainfall of 30 mm or more in 24 hours is recorded and ground displacement is equal to or is more than 2 mm in 24 hours. Rainfall will be measured by the representatives of the inhabitants and Ministry of Local Government. As soon as a 30mm/ 24hrs is recorded the information will be communicated to the Director, Meteorological Services. The Meteorological Services will confirm the recording and transmit it to the Police Department–Information Room–which in turn will communicate it to the Prime Minister's Office, Ministry of Local Government.

On being informed that the 30 mm rainfall in 24 hours has been recorded, the Ministry of Local Government will start taking daily readings of extensometers to measure ground displacement. On 2 mm a day or more displacement being recorded, the Ministry of Local Government will communicate the reading to the Meteorological Services, Police Department and the appropriate local Authorities. The Chairman of the Coordinating Committee will then confirm that Stage I has been reached.

The Stage 1 Warning will be communicated to the representatives of affected inhabitants by Police Department. The warning will also be communicated by Police–Information Room–to the following Ministries and Organisations which make up the Coordinating Committee.

☆ Prime Minister's Office

☆ Ministry of Local Government

☆ Ministry of Public Infrastructure, Land Transport and Shipping

☆ Ministry of Health and Quality of Life

☆ Ministry of Social Security, National Solidarity and Senior Citizens Welfare and Reform Institutions

☆ Ministry of Education and Human Resources

☆ Police Department

☆ Ministry of Public Utilities

☆ Meteorological Services

☆ Government Fire Services

☆ Central Water Authority

☆ Central Electricity Board

☆ Local Authorities and

☆ Universities

When issuing the Stage 1 warning the Police will advise the inhabitants of landslide prone areas through their representatives, to start preparing themselves to move out of their houses in accordance with instructions already issued to them.

Stage 2: Warning Stage

The Stage 2 Warning will be triggered on a ground displacement of 1 cm in 24 hrs being recorded. The Ministry of Local Government will monitor constantly the ground movement and will inform the Meteorological Services and the Police Department as soon as a displacement of 1 cm / 24 hours has been recorded. The Police Department will inform PMO and the Chairman of the Coordinating Committee will convene a meeting for the issue of the warning No 2. The warning will be issued for broadcast by the MBC and private radios. Communication to the affected residents will be done by the Police Department and Fire Services by loudspeakers or other means.

The MBC and / or Police Department and Fire Services shall, when issuing Stage 2 Warning advise the residents to complete all preparations for an eventual evacuation and stand ready to vacate their houses once the order is issued. Arrangement should be made by the Ministry of Health for the transfer to hospitals of disabled people who elect to do so. First Aid Service providers may assist.

The Police Department will also contact the PMO and Ministry of Local Government with a view to convening at the earliest a meeting of the Crisis Committee consisting of representatives of:

☆ Prime Minister's Office

☆ Ministry of Local Government

☆ MPI

☆ Police Department

☆ The Meteorological Services

☆ Local Authorities

☆ Ministry of Public Utilities

The Crisis Committee will review the situation in the light of all available information pertaining to rainfall recording and ground displacement. The Police Department will take appropriate measures to muster all available resources and equipment in order to assist in an eventual evacuation exercise and any salvage operation. The Stage 2 Warning will also be communicated by the Police Department to the following Ministries/ Departments and Organisations and which will be responsible for the following:

Ministry of Health

1. To prepare special ward for any casualty that may arise out of an eventual evacuation;
2. To provide an adequate number of medical and paramedical personnel intended to receive casualties; and
3. To be ready to despatch Ambulances adequately staffed and equipped.

Ministry of Social Security

1. To ensure that personnel are ready to proceed to appropriate Refugee Centres as and when the evacuation order is issued in order to attend to the evacuees; and
2. To arrange for the provision of basic necessities to evacuees.

The Central Water Authority

The Central Water Authority will stand ready to close the shut-off valves on the pipes going through the region as soon as the evacuation order is issued.

The Central Electricity Board

The Central Electricity Board will be ready to switch off electricity supply in the affected areas as and when instructed by the Crisis Committee or the most senior gazetted Police Officer. The Central Electricity Board will ensure as far as possible that power cuts are restricted to the affected areas only so as to avoid unnecessary deprivation of electricity to unaffected areas.

Ministry of Local Government

Ministry of Local Government will continue to take readings of extensometers as frequently as may be appropriate to determine whether the ground displacement progresses beyond 1 cm/ 24 hours and ensure that the information is communicated to the Police Information Room and the Meteorological Services.

The Fire Services and the Non-Governmental Organisations

The Fire Services and the Non-Governmental Organisations (Red Cross Society, St. John Ambulances etc.) will be informed by the Police of the possibility of an evacuation order being issued and their assistance enlisted.

Stage 3: Evacuation Stage

Stage 3 is reached when ground displacement is equal to or greater than 2mm in an hour and rainfall will continue. As in the case of the two previous stages the reading of the extensometers will be continually monitored by the Ministry of Local Government and the data communicated regularly to the Police Department, and Meteorological Services. The Police will then pass on the information to the Crisis Committee which will meet to approve the evacuation order. The evacuation order will be broadcast and/or communicated to the appropriate residents in the same manner as in Stage 2.

If, on information being obtained from the Ministry of Local Government, the Police Department considers that an urgent and immediate evacuation is required and that there might not be enough time to convene the Crisis Committee then the most senior gazetted officer will give the order for evacuation after consultation with the Chairman of the Crisis Committee, if possible.

As Stage 3 is reached and evacuation is in progress the various Ministries/ Departments/ Organisations involved should actively set in motion arrangements for which they are responsible. In particular the following measures should be implemented.

1. *The Ministry of Education*: Schools in affected areas should be closed;
2. *Ministry of Social Security*: Staff should be in attendance where appropriate. The Ministry will also make arrangements for the provision of basic necessities to the evacuees;
3. *Ministry of Health*: Ambulances should be despatched on site for the conveyance of handicapped, old and sick people, and any casualty cases

to hospital. Arrangements will also be made for Health Inspectors to visit regularly the refugee centres to ensure acceptable sanitary conditions there;

4. *Central Water Authority*: The Central Water Authority will close the valves on the pipelines within the affected area and will arrange for water to be supplied regularly to the refugee centres;

5. *Central Electricity Board*: The Central Electricity Board will proceed with the interruption of the power supply in the affected areas;

6. *Police Department* will cordon off the affected area and ensure the protection of property of the residents. An Incident Officer will be responsible and have full authority for the control and coordination of the operation on site. Access to the cordoned off area will only be permitted by the Incident Officer.

Stage 4: Emergency Stage

When there is sudden landslide and the Crisis Committee cannot for practical reasons be convened, the Emergency Warning is issued by the Police department after consultation with the Chairman of the Crisis Committee, if possible. Action will be triggered off as provided for under Stage 3: Special Arrangements during cyclone warnings/ torrential rains warning.

The prevalence of cyclonic conditions over and around Mauritius will entail the adoption of special arrangements with regard to the inhabitants of landslide prone areas. The issue of a Cyclone Warning Class 2/ torrential rain warning may constitute for the inhabitants a Stage 2 Warning. Being given that the issue of a Cyclone Warning Class 3 entails the cessation of all normal activities the inhabitants of these areas may be evacuated if there exists a strong likelihood of a Cyclone Warning Class 3 being issued.

Therefore, as soon as a cyclone warning class 2 or torrential rain warning is issued by Meteorological Services the Police department will in consultation with the Chairman of the coordination committee, convene a meeting to consider the advisability of issuing an evacuation order. Action as provided in Stage 3 will be triggered off.

Stage 5: Termination

A close monitoring and stocktaking exercise will be undertaken. After stabilisation of ground movement has been noted, the all-clear signal will be given after a meeting of the Crisis Committee.

Public Awareness Campaign

Early Warning Systems can only be effective if they are easily understood by the general public and important stakeholders. Awareness campaign is becoming more and more important and the government is encouraging all institutions, especially the Meteorological Services, to undertake aggressive public awareness campaign in order to sensitize the population, at the grassroot level, to understand the importance of taking all precautions against the adverse effects of natural hazards.

A concrete and solid awareness and preparedness action plan has been put into place. Both the electronic and print media are being requested to play an important role in the dissemination of awareness campaign. The press has a wide diversity of audience and is therefore an effective means of informing and educating the public. Regular press releases and reports on adverse weather conditions and their impacts are given wide publicity through information exchange between the Meteorological Services and the media. Special efforts have been made to reach schools and other educational institutions. "Open house" are organized regularly during special events such as the World Meteorological Day or World Water Day to bring thousands of visitors to become familiar with early warning systems through exhibits.

Special brochures and posters are distributed freely to members of the public. One-day workshops are organized to sensitise stakeholders on ways and means to protect life and property. A better prepared community will ultimately help to reduce the vulnerability of human life and property and also to mitigate the adverse effects of natural hazards.

Conclusion

With the assistance of the government, the National Meteorological Services is bringing together stakeholders from a wide range of users of early warnings, which are people- centred and are based on the needs, priorities, capacities and cultures of those at risk. People are being encouraged to be partners in the system, not to be controlled by it.

References

[1]   Air Mauritius (1994) Operations Manual, Vol I Part II, Cyclone Operations,18/12/94, Mauritius.

[2]   Budgen, P and Bettany, B W (1996) Report of Visit to Mauritius 12-17 May 1996, Meteorological Office, Bracknell.

[3]   Central Electricity Board (1993) Operating Instructions for Cyclone Conditions, Internal Regulations No.7, General Staff Instructions Circular No. 15/19/12/93, 17.12.93, Curepipe.

[4]   Dukhira, C G (1994) Grass Roots Democracy for National Development, Editions de l'Ocean Indien, Mauritius.

[5]   Mauritius Expert Processing Zone Association (1995) Annual Report 1995, Port Louis.

[6]   Mauritius Meteorological Service and Ministry of Information (1994) Cyclone Watch, Republic of Mauritius.

[7]   Meteorological Department(1996) Annual Report of the Meteorological Department July 1991 to June 1994, Mauritius.

[8]   Mileti, D (1974) Natural Hazard systems in the United States: A research assessment, Institute of Behavioral Science, University of Colorado, Boulder, Colorado.

[9] Padya, B M (1984) The Climate of Mauritius, Meteorological Office, Mauritius.

[10] Padya, B M (1989) Weather and Climate of Mauritius, Mahatma Gandhi Institute Press, Moka, Mauritius.

[11] Parker, D J and Fordham, M (1996) An Evaluation of Flood Forecasting, Warning and Response Systems in the European Union, Water Resources Management, 10, 279-302.

[12] Prime Minister's Office (1995) Disaster Preparedness, Cyclone and Other Natural Disasters Scheme, Port Louis.

[13] Republic of Mauritius, Central Statistical Office (2005) Digest of Agricultural Statistics, Port Louis.

[14] Republic of Mauritius, Central Statistical Office (2005) Digest of Demographic Statistics, Port Louis.

[15] Republic of Mauritius, Ministry of Housing, Lands and Town and Country Planning (1995) National Physical Development Plan, Port Louis.

[16] United Nations, Department of Humanitarian Affairs (1994) Cyclone "Hollanda" and Cyclone "Ivy", Report on the Assessment Mission to Mauritius and Rodrigues Island, 18-27 February 1994, DHA, Geneva.

[17] Walker, I (1993) The Complete Guide to the Southwest Indian Ocean, Cornelius Books, Argeles sur Mer, France.

[18] World Meteorological Organisation (1989) Human Response to Tropical Cyclone Warnings and their Content, WMO Technical Document Report TCP-11, Tropical Cyclone Programme, Geneva.

Chapter 3

Policy Issues and Mitigation Strategies of Natural Disasters Management in Bangladesh

Munir Ahmed
*Lightning Awareness Center, and
TARA-Technological Assistance for Rural Advancement,
1 Purbachal Road, Northeast Badda, Dhaka 1212, Bangladesh
E-mail: tara@citechco.net, munir_tara@yahoo.com, tara@ranksitt.com*

ABSTRACT

Bangladesh is country of highly prone to natural disasters and almost all the disasters roam the country now and then. Flood and cyclone are the daily-company and occur almost every year in the country. The natural disasters cause severe damages of lives and assets is beyond the capacity of recovery. The country is slowly progressing towards mitigation and trying to adopt long-term mitigation policy.

After the floods 1980s and cyclone of 1991, the concept of disaster management acting only after the occurrence has been replaced by the concept of total disaster management involving prevention/ mitigation, preparedness, response and recovery. The government is quite committed. The Disaster Management Bureau of the Ministry of Food and Disaster Management has been proactive and taken initiatives. As a part of its initiative, it has prepared Policy and strategies for long-term mitigation. A strategic plan has been prepared and in implementation. The standing order has been released making understand everyone's responsibility. Committees have been formed from national level to grass-root level. Six areas have been identified as the focus areas for effective and sustainable mitigation. Importance has been given on community awareness and responding to it. There has been support from NGOs and international community in the process. As a part of management, the Government has constructed good number of structures like shelter houses, embankments as preventive measure and tackling post-disaster situation. The overall progress and steps appears to be enthusiastic.

Keywords: Disaster, Bangladesh, Flood, Cyclone, Tornado, Lightning, Strategy, Policy, Mitigation.

Introduction

Bangladesh is one of the most disaster prone countries in the world. Cyclones, devastating floods, riverbank erosion, drought, earthquake, arsenic contamination, chemical pollutants, fire and roadside accidents are major incidents that leave negative consequences.

The conventional disaster management model (response, relief and recovery) is being replaced by internationally with a more holistic model. Under this holistic approach process of hazard identification and mitigation, community preparedness, integrated response efforts and recovery are planned for and undertaken contiguously within a risk management context to address issues of vulnerability.

Disaster Management Bureau of Bangladesh has a strategic plan. Main objective of this plan is to provide a statement of key strategies for the provision of services that align with the Ministry of Food and Disaster Management Corporate Plan 2005-2009.

Natural Disasters in Bangladesh

It has been mentioned earlier that Bangladesh is very susceptible to Natural Disasters. The kinds and extent of disasters are also very high. Probably there is no natural disaster on the universe which does not occur in Bangladesh. The disasters causes irreparable losses of lives and properties most of the years. The poor country is incapable in handling the losses. Some major natural disasters are briefly described in the following:

Flood

Bangladesh is probably the most victimized country in the world by flood. It occurs through continuous and excessive rainfall, the flowing of water from the upstream, the decrease of water-depth in the river and other causes the flood in the country. It does occur in monsoon (June-August). The flashflood comes in April-May and sometimes there is flood in late monsoon during September-October. The extent of flood is virtually all over the country except Chittagong Hill Tracts and other high areas. The flood damages the earthen houses, standing crops, land erosion, livestock and human life damage. The losses become very high when the flood damages the embankment. The people feel safe being in the protected area and they don't have any preparation for it. The flood of 1988 during August-September inundated 89,000 sq. km and caused loss of 1,517 human lives. The 1998 flood was within a very long duration of 65 days and inundated about 100,000 sq. km and took 918 lives. The 2004 flood in Bangladesh inundated 40 districts and it took lives of 747 people. The other severe floods in recent years were 1974, 1987, 2002 and 2004.

Quite good effort has been put forwards and many interventions have been taken. However, because of its geographical location, the threat of flood can't be root-out completely. The government has constructed Flood Control and Drainage (FCD), Flood Control Drainage and Irrigation (FCDI), Coastal Poldering, Dyke on River Bank (*e.g.* BRE), Sluice gates/regulators, flood proofing (homestead raising), submersible roads, drainage system improvement, dredging of rivers etc. It should be mentioned that the embankment/FCD/FCDI could not be extended to all areas and rivers due to financial lacking and environmental issues. There are some coping mechanisms in established like shelter house, training to people, homestead raising etc. The combined effort of government, NGOs, civil society, national and international donors is required to compete with the flood disaster.

Cyclone

Cyclone originates from the depression in the Bay of Bengal as strong storm sometimes accompanied by water rise. The coastal and bay areas are prone to the cyclone. Usually the cyclone occurs during September-October in the country. It is one major natural disaster in the country which takes lives and damages assets. People are helpless to the high strength of cyclone. When water-rise accompanies the cyclone, the losses become higher. The severe cyclone of 12 November 1970 took 0.3 million human lives in Bangladesh and damaged the property of billions dollars. Another worst cyclone on April 1991 killed 0.14 million people. The cyclones of 1876, 1919, 1961, 1963, 1965, 1970, 1985, 1988, 1991, 1994, 1995 and 1997 were also severe. The government has some established cyclone centers in the prone area. NGOs and donors are assisting to the government. Effective and modern cyclone-warning mechanism and awareness raising among people need to be established along with cyclone centres.

Tornado

Tornado is the whirling air along with thunder which hits on land like the elephant's trunk. It occurs so suddenly leaving no chance for forecast and it lasted for a very short period. It occurs during March-May. It occurs all over the country especially the dry areas are more vulnerable. It does take off houses, trees, electric poles and any firm structures with its mighty power destroying properties and lives. Sometimes, the losses become very high.

Lightning

Lightning is the electric shock that occurs from the collision of the clouds with thunder. In Bangladesh, it usually occurs between March and June and throughout the whole country. People and animal dies, trees and infrastructures are damaged by the lightening. Effective mitigation measures yet to be established. Lightning Safety Awareness programme is needed to reduce loss but it is being done in very limited form. Lightning Awareness Centre (LAC) at TARA NGO and Bangladesh Lightning Research Centre at Jahangirnagar University are working closely for better mitigation. According to the LAC-TARA data, a total of 127 people died and 78 injuries in 2006 from lighting.

Earthquake

Earthquake is the trembling or shaking movement of the earth's surface. Bangladesh is surrounded by the regions of high seismicity and prone to earthquake. Although no larger earthquakes and devastating damage has occurred yet but the country has experienced many minor tremors. The earthquake of 1869, 1885, 1897, 1918, 1930, 1934, 1959, 1997 and 1999 are noteworthy. The country is at the preliminary stage of earthquake mitigation measures.

Drought

Drought is another severe natural disaster which shows up at some intervals in Bangladesh and occurs from scarcity of rain. It occurs during March-July and most of the area of the country. It causes disastrous crop failures causing famine. In 1979, the country experienced a severe drought, which was termed as the worst in the recent past. The droughts of 1957 and 1972 were also severe nature. Some initiations have taken to address the issue. Generally major part of North West region of Bangladesh is prone to draught and a programme has been developed to cope with the situation.

Famine

A state of extreme starvation suffered by the population of an area due to scarce food supply is the famine. In local language, it is termed as "Monga". Drought and other reasons contribute to famine. The northern part of the country is prone to it and usually happens March-June. The famine of 1866, 1896, 1943 and 1974 are remarkable. The government and NGO's are working on reducing the famine by keeping food reserves and creating IGA.

Water Logging

The permanent/longer period accumulation of water in cultivatable land making them unusable for production is the state of water logging. It is becoming an emerging problem in the country. The construction of unplanned FCD/FCDI, roads, embankments, dams, infrastructure, enclosures, and fish and shrimp farms causes the water logging. There are many areas of south-west of the country is affected by it. The water logging of "Bhabadaha" in Jessore has caused the displacement of thousands households, huge area has become unproductive and drinking water scarcity. Steps are being taken by improving the drainage system, constructing sluice gates and digging canals.

Tsunami

A tsunami is a huge ocean wave that travels at high speeds (up to 965 km/hr), hundreds of miles over open sea before it hits land. A tsunami is usually caused by an earthquake, volcanic eruption or coastal landslide. Bangladesh is on the borderline of Tsunami prone area and has chances to be affected. The Bay of Bengal is such area. So far no Tsunami has attacked in the country but the Tsunami in Sri Lanka alarmed the country. During that time high water level hit the coast line but no news of death received. Step yet to be taken to address the issue.

Issue in Disaster Management

Though Bangladesh wishes to have a new focus on disaster management but it has many challenges which are as follows:

☆ Lack of capacity to cope with unprecedented change in technology;

☆ Lack of timely need to embrace a knowledge management and knowledge and information sharing culture;

☆ Population demographics are to a changing towards urbanization and industrialization compared to rural in earlier time;

☆ Need to develop trained workforce and also to retain that trained skilled force, capable in delivering our new services and programmes related to disaster management;

☆ Need for staff in Disaster Management Bureau and Non-Governmental Organizations that would be capable in adapting in changed situations;

☆ Need to further development of disaster management programmes and services that recognize issues of gender and the socially disadvantaged people;

☆ Need to understand the full impact of climate change on the environment and to consider activities in disaster management accordingly;

☆ Need to consider disaster management more wide and long term perspective than mere relief work (disaster management from conventional response and recovery). In fact disaster management is to be a part of environment management plan (EMP).

Future Expectations towards Mitigation Measures Related to Policy

Bangladesh has to go a long way for the mitigation of the natural disaster. A proper policy and its implementation is a must. The expectations in this regard are quite high for the future. The expectations are:

☆ To strongly advocate the adoption of comprehensive approach to risk reduction and risk management, that is based on the international best practice model (AS/NZ 4360-1999).

☆ To engage the whole of Government system in risk reduction and risk management through mainstreaming and advocacy strategies.

☆ To develop policy and operational frameworks for sustainable coordination, collaboration and information management across Government, and with key disaster management stakeholders.

☆ To establish formal partnership with Government agencies, NGO, civil society and the private sector for effective and sustainable service delivery.

☆ To have a community focus with a strong emphasis on issues of gender and socially disadvantaged embedded in programmes.

☆ To establish and maintain strong regional partners and networks, and to actively contribute to national, regional and international disaster management agendas.

☆ To pursue the standardization of disaster management training and systems to align with best practice standards, models and competencies.

☆ To build and maintain the professional competencies of related staff engaged in disaster management.

Policy (Standing Order) in Disaster Management in Bangladesh

The GoB has prepared the Standing Orders with the declared objective of making the concerned persons understand their duties and responsibilities regarding disaster management at all levels, and accomplishing them. All ministries, Divisions/ Departments and Agencies shall prepare their own Action Plans in respect of their responsibilities under the Standing Orders for efficient implementation. The national Disaster Management Council (NDMC) and Inter-Ministerial Disaster Management Coordination Committee (IMDMCC) will ensure coordination of disaster related activities at the national level. Coordination at district, Upazila (sub-district) and Union level will be done by respective District, Upazila and Union Disaster Management Committees. The DMC will render all assistance to them by facilitating the process.

The Ministries, Divisions/Departments and Agencies will organize proper training to their officers and staff employed at District, Upazila, Union and village levels according to their own Action Plans so that they can help in rescue, evacuation and relief work at different stages of disasters. The local Authority shall arrange preparedness for emergency steps to meet the disaster and to mitigate distress without waiting for government help. The Standing Order should be followed during Normal times, Precautionary and Warning stage, Disaster stage and Post-disaster stage.

Disaster Management/Mitigation Strategy

After the floods of late 1980s and the killer cyclone of 1991, the concept of acting only after the occurrence of disaster has been replaced by the concept of total disaster management involving prevention/mitigation, preparedness, response, recovery and development. The GoB has, therefore, total commitment towards reduction of human, economic and environmental costs of disasters by enhancing overall disaster management capacity.

Efforts have been continuing for optimum coordination and best utilization of resources along-with ensuring community involvement so that they are aware of what they can do for protecting their lives and properties against disasters. The plan and conduct of disaster management by GoB involve preparedness, response, recovery and mitigation as key notes for building up self-reliance of the community people. The DMB has prepared Strategic Plan based on the strategy. The strategic plan is to provide a statement of key strategies for the provision of services that align with the Ministry of Food and Disaster Management's Corporate Plan 2005-09.

The Government Vision towards mitigation strategies is "to reduce the vulnerability of the people, especially the poor, to the effect of natural, environmental and human induced hazards to a manageable and acceptable humanitarian level". The Mission of MoFDM is "to achieve a paradigm shift in national disaster management strategies from conventional response and recovery to a more comprehensive risk reduction culture, and to promote food security as an important factor in ensuring the resilience of communities to hazards". The objective of the MoFDM is "to strengthen the capacity of the Bangladesh disaster management system to reduce unacceptable risk, to improve response and recovery management at all levels and to effectively integrate and manage the national food security system". The role of Disaster Management Bureau in this regard is "to provide effective, contemporary and professional disaster risk management services, specializing in risk management, mitigation, awareness, education, information and response and recovery coordination services".

Supporting Partners of DMB

The DMB has the following partners in disaster mitigation:

☆ Ministries of the Government of Bangladesh;

☆ Non Government Organizations (NGO);

☆ Civil Society;

☆ Business sector.

Key Services to Disaster Mitigation by DMB

DMB offers following key services:

☆ Government and Non-Governmental Organization coordination;

☆ Training;

☆ Community Education and Awareness;

☆ Media and Public Information;

☆ Dissemination of new information relating to disaster and preserving information;

☆ Emergency Response and Recovery Coordination;

☆ Disaster Risk Management Advise;

☆ Other specific duties as assigned in the Allocation of Business.

Disaster Management Mechanism

The GoB has formulated a set of mechanisms to maintain proper co-ordination amongst the concerned Ministries, organizations and line agencies and also to ensure their effective functioning during emergency. For the mechanisms to be operative, a guidebook named 'Standing Orders on Disaster' has been published as a basic tool which has been mentioned earlier.

The Standing Orders outline the activities of each Ministry, major agencies / Departments so as to handle emergency situations efficiently. In the efforts of making the mechanisms clear and comprehensive, National Policy on Disaster Management has been designed and under process for final approval.

The initial operational direction and co-ordination for any disaster situation come from the highest level of institutional arrangement (*I.e.;* NDMC) through second highest authority IMDMCC for overall disaster management in the country. Committees from National to Grassroots levels (*i.e.* DDMC, UzDMC and UDMC) under the framework of disaster management guidebook *i.e.* Standing Orders on Disaster work on Disaster Management.

Disaster Management Model-Bangladesh

The model that has been adopted in Bangladesh for disaster Management is shown in Figure 3.1.

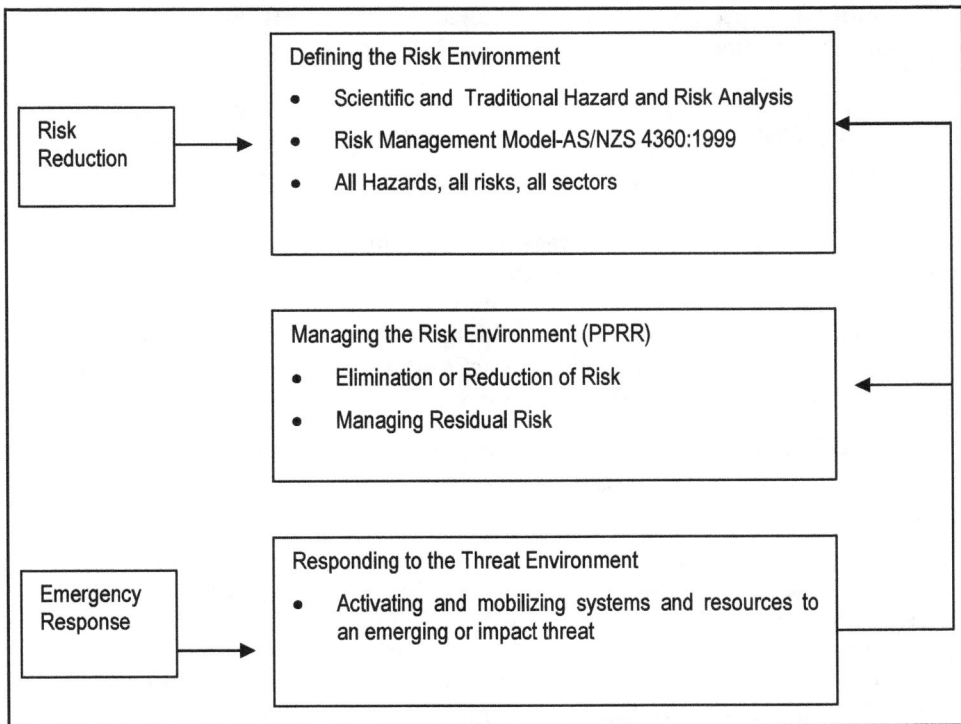

Figure 3.1: Disaster Management Model of Bangladesh.

Key Strategies for Disaster Management

Six key strategies for disaster management have been adopted by the DMB for the country. They are as follows:

Strengthening and Professionalizing the Disaster Risk Reduction System

The Goal of the strengthening and professionalizing the disaster risk reduction system is to establish a professional, capable and skilled best practice Disaster Management System. The strategies in this regards are the development and implementation of a business planning process, including the development of a DMB Strategic Plan and annual operational plans. It plans to develop and implement a Professional Development Strategy for DMB staff, ensuring all staff undertakes training and development activities consistent with their role and responsibilities. The development and implementation of a Disaster Management Profile Enhancement is also a part. The enhancement strategy will include i) partnership development with NGOs, civil society, training and academic institution and other Ministries ii) frameworks for joint programmes across Ministries and with NGO and others and iii) a review and refocus of various disaster management groups and organizational programmes. This strategy covers ensuring DMB services are suitable to the role and responsibilities of the Bureau and the developing a national legislative and policy framework. The framework will includes i) the preparation and promulgation of the draft Disaster Management Act ii) the development of a National Disaster Management Plan iii) the development of a National Disaster Risk Management Policy and iv) the review, revision and reproduction of the Standing Orders for Disaster. The last strategy of the strengthening and professionalizing is the revise and align the DMB organizational structure consistent with the new MoFDM Allocation of Business and the requirements to implement the strategies contained within this strategic plan.

The key outcomes of this strategy will be (*i*) professional and competent staff with a greater knowledge and understanding of comprehensive disaster management; (*ii*) increased and more effective information sharing and coordination within and across government, NGOs, and other stakeholders; (*iii*) an established legal and policy framework based on recognized best practice disaster management concepts and (*iv*) an effective, well structured and resourced, high achieving.

Mainstreaming of Disaster Risk Management Programming Through Coordination, Co-operation and Advocacy (Partnership Development)

The goal is the mainstreaming of Disaster Risk Management principles and practices in the whole-of-government and national development processes. The strategies in the mainstreaming would be the managing and creating an active committee system at the national level. The national committee will support to some other committees like National Disaster Management Committee, Inter-Ministerial Disaster Management Co-ordination Committee and National Disaster management Advisory Committee. The strategy will refocus on the activities of disaster management committee from conventional response and recovery to a more comprehensive risk management approach. It will review, rationalize and develop a system of effective advisory and supporting committees including committees for which DMB has direct responsibility. The professionalizing and creation of an international best practice disaster risk management training system in the work plan. It will include: the development and implementation of a national Disaster Management Training Policy, the review and revision of a National Disaster Management Training Plan consistent with the policy and curriculum, and the establishment of a National Disaster Management Training Centre (Institute of Disaster Risk Management and Research).

The key outcomes of the mainstreaming of disaster risk management will be: the disaster management committees at all level will have increased capacity to actively engage in risk reduction and emergency response activities. The committee meetings will be conducted regularly at national level and will be effective in engaging all Ministries in disaster risk management.

Strengthening of Community Institutional Mechanisms (Community Empowerment)

This will be another strategy with the goal of developing of disaster resilient communities that will have enhanced coping capacities across a broader range of hazards. The strategy towards the goal will include the development and implementation of education and awareness programmes that are aligned with identified risk treatment strategies. Such strategy will include the publishing of an Indigenous Coping Mechanism Guidebook and creating and developing of innovative awareness and community education programmes to include the integration of comprehensive disaster management concepts within the education system. The support for the development and introduction of methodologies for integrating disaster management planning at a local level within development planning process will an element of the strategy. The strategy will support the development and implementation of the Local Disaster Risk Reduction Fund in community reduction programmes. It will also analyze and integrate the issues of Gender and Social Exclusion into DMB policies, programmes and services.

The key outcomes of the strengthening of community institution are expected that the community risk reduction programmes will have a strong focus on issues of gender and the socially disadvantaged. It will increase community understanding of local risks and potential risk treatments related to their treats. The frameworks to facilitate the integration of community risk assessment outcomes within government agency development plans are established and operational and the Local Disaster Risk Reduction Fund contributing to community risk reduction activities.

Expanding Mitigation, preparedness and response strategies across a broader range of hazards

This strategy area will focus on the development of policies and programmes that recognize climate change impacts on hazards and risks to communities, and mitigation strategies that are based on a risk management assessment. The adopted strategy will be the adequate addressing of the assessment of Bureau programmes and services to ensure climate change issues create awareness within National Disaster Management committees on climate change issues and to support the development and implementation of Community Risk Assessment Guidelines. The guidelines will include, but not limited to Earthquake, Tsunami, Rural Flood, Urban Flood, Flash Flood, Cyclone, Tornado, River Bank Erosion and Environmental Pollution. The strategy will also support the development of a Drought Prediction Model and its integration into community risk assessment processes, support the Bangladesh Fire Services and Civil Defence (BFSCD) in the analysis and assessment of needs and priorities for an Urban Search and Emergency Capability and support the Ministry of the Environment and Forest (MoEF) in the integration of climate change research within community risk assessment and emergency response models.

The expected key outcomes are (*i*) an information database and climate change research that informs comprehensive disaster management programmes and response; (*ii*) standardized risk assessment guidelines that incorporate climate change and formal hazard and risk analysis considerations and (*iii*) a plan to enhance the Urban Search and Rescue capability developed and approved.

Strengthening Emergency Response and Recovery (Relief and Rehabilitation) Systems (Operationalizing Response and Recovery)

The goal of this key result area is to create greater levels of coordination and cooperation at regional and national levels, and enhanced whole-of-government information, warning and reporting systems. The strategy in this regard will include i) develop, validate and implement preparedness, response and relief management systems based on the (*i*) All Hazards model; (*ii*) create and develop an effective emergency response information/situation reporting system from the national and community level; (*iii*) develop frameworks and policies for the integration of NGOs, civil society, private business, and national and international support into the emergency response and recovery planning and coordination processes and programmes and (*iv*) develop and maintain an effective emergency response and recovery coordination system. The coordination system will include (*a*) the establishment and operationilization of a National Disaster Management Information Centre; (*b*) the establishment and operationilization of District Disaster Management Information Centres; (*c*) the development of a Damage and Needs Assessment Methodology and its implementation within response planning systems; (*d*) the integration of a Food Security Policy within relief operational procedures; (*e*) the development and implementation of a Search and Rescue capacity building programme and (*f*) the development of an All Hazards community warning system.

The outcomes of the Strengthening emergency response and recovery (Relief and Rehabilitation) Systems are expected to have (*i*) an effective Disaster Information Management and Coordination Centre at the national level, and targeted districts; (*ii*) an effective national and community all hazards warning system; (*iii*) greater levels of coordination and information acquisition and dissemination across government agencies, NGOs, and civil society networks; (*iv*) enhanced national and regional cooperation and networks; (*v*) Coordinated, timely and appropriate response and more effective damage assessment, relief and recovery system and (*vi*) urban search and rescue capability progressively enhanced and tested.

Supporting and Complementing the National Food Security System

This strategy intends to support and complement the development of a robust, well managed, equitable and disaster risk resilient national food security system. The strategy are (*i*) support the development, strengthening and implementation of a national food security policy and strategies; (*ii*) assist in the establishment, maintenance and enhancement of a dependable national food security system; (*iii*) provide input into food planning, research and monitoring and (*iv*) support DGoF (Directorate-General of

Foods) in the management of drought, famine and food storage situation. The key outcome of this strategy is the support to the DgoF is timely and effective.

Major Programmes of Disaster Management in Bangladesh

The government of Bangladesh has taken programmes for the disaster management of the country from its own resources along with the support from NGOs, civil society and donor countries. The existing system for disaster management in the country covers activities at normal times for important disaster management aspects like mitigation/prevention, preparedness, response and recovery. The Government as part of disaster management, has been trying to develop its scientific networking in respect of disaster forecasting and early warning. The management covers both the structural and non-structural mitigation. Some disaster management practices are mentioned in the following:

Cyclone Preparedness Programme

As part of structural mitigation measures, GoB has so far constructed 2,023 cyclone shelters/ shelter houses. About 3,931 km. long coastal embankment has been constructed to protect coastal land from inundation by tidal waves and storm-surges. With the grant from the Japan Government, the radar at Agargaon, Dhaka has been replaced. New radar at Rangpur and a satellite ground receiving station at SWC, Dhaka has been established in 2005. It has enhanced the capability of Bangladesh Meteorological Department in cyclone and flood forecasting.

DMB with the funds and assistance from donor and NGOs has established councils and committees from central level to local level. These are National DM council, Inter Ministerial DM Council, National DM Advisory Committee, DMTTF, FPCG, NGOCC, CSDDWS, District DM Committees, Upazilla DM Committees and Union DM Committees. In costal areas, there are disaster preparedness committees at Village and Upazila levels who meets at regular interval and frequently during cyclone or other disaster time. The information passes quickly to all people of the area by the village committee representatives.

Flood Preparedness Programme

The Government has established 200 flood shelters across the country. Many FCD/FCDI have been established. A total of 4,774 km. (in length) drainage channels have been constructed so far. Many school yards of flood prone area have been raised which are used by flood affected people. Some NGOs, projects and donors are contributing in flood mitigation. CARE has built wave protection and erosion protection walls in the Haor area to reduce the force of wave and protect the common homestead area from erosion. The CLP (Chars Livelihoods Programme) is raising individual homestead in char areas and CONCERN has some activities to this regard. As non-structural mitigation, substantial progress in the expansion of flood forecasting and warning services (FAP-10) in the country has been made with the help of Danish Hydraulic Institute.

Earthquake Preparedness Programme

It is on its early stage. Some surveys are on progress by NGOs and donors.

Lighting Preparedness (Safety Awareness) Programme

LAC (Lightning Awareness Center) at TARA (Technological Assistance for Rural Advancement) has been working in this part and there has been good progress. Several workshops have been organized in the country. Awareness programme has been conducted at community level. A poster has been developed which has been very effective measure for disseminating the lighting awareness messages. LAC of TARA has good linkages in other such centers in South Asia.

Drought and Famine Preparedness Programme

The government and some NGOs are working to tackle the drought and famine situation. The government has special allocation and storage of food in this purpose. The NGOs are creating income generating activities for the affected people.

Other Non-Structural Mitigation

Legislation and Policy

Disaster Management Legislation has been drafted with the purpose of providing for the formulation of disaster management policy relating to preparedness and emergency measures, and rehabilitation programme to deal with disaster.

Training and Public Awareness

☆ As part of training and public awareness nearly 50,000 people related to disaster have been trained through 500 courses/workshops/seminars symposia.

☆ As part of public awareness activities, booklets containing information about cyclone, flood, tsunamis etc. and calendar, posters depicting disaster points have been regularly printed and distributed up to the grass-root levels.

☆ To raise awareness among the students on various hazards/disaster management, a chapter on disaster management has been included in the educational curricula from classes V to XII.

☆ GoB has decided to make compulsory a session of at least 02 hours on disaster management in the training curricula of all types of Training Institutes to train officials and non officials.

Conclusion

Bangladesh has got quite improvement on natural disaster mitigation. The Government and has taken good initiatives through the Ministry of Food and Disaster Management and Disaster Management Bureau. The good thing is that there has substantial progress on developing Policy and Strategies and both are in place. There is an action/strategic plan by DMB up to 2009 prepared on based on the policy and strategy. The implementation of the plan is running. It is hoped that Bangladesh will be efficient enough on complete implementation of its policy and strategy in the near future. The Government needs the updating and revising (if necessary) to fit in any emerging situation. The overall success will depends on the strong will of the

Government and its line agencies. Support from NGOs, Civil Society and International Community is also essential.

Acknowledgements

The author has a very highly debt to the Disaster Management Bureau (DMB) of the Bangladesh Government. It has provided him support with information and documents which has been very useful for preparing this paper. It should also be acknowledged that much of this document has been taken from the website and the strategic plan of the DMB.

The author would like to express sincere thanks and gratitude to Dr. Chandima Gomes of University of Colombo, Sri Lanka and Mr. Richard Kithil, NLSI, USA for their cooperation and information especially on Lightning Safety part.

References

[1]   *Disaster Management Bureau, 2005:* Turning Strategies into Action: Strategic plan 2005-06.

[2]   www.dmb.gov.bd: The official website of the Disaster Management Bureau, Bangladesh.

Abbreviations

BFSCD: Bangladesh Fire Services and Civil Defence

BLRC: Bangladesh Lighting Research Center

BRE: Bhramputra Right Embankment Project

CLP: Chars Livelihoods Programme

DDMC: District Disaster Management Committee

DMB: Disaster Management Bureau

EMP: Environmental Management Plan

FAP: Flood Action Plan

FCD: Flood Control and Drainage

FCDI: Flood Control Drainage and Irrigation

GoB: Government of Bangladesh

IMDMCC: Inter-Ministerial Disaster Management Coordination Committee

LAC: Lighting Awareness Centre

MoEF: Ministry of the Environment and Forests

MoFDM: Ministry of Food and Disaster Management

NDMAC: National Disaster Management Advisory Committee

UDMC: Union Disaster Management Committee

UzDMC: Upazila Disaster Management Committee

Chapter 4

Recent Developments in Disaster Management in India

Vinod K. Sharma
Disaster Management, Indian Institute of Public Administration,
New Delhi, India
E-mail: profvinod@gmail.com

ABSTRACT

In view of the frequency of natural disasters in the country, a well-structured and integrated disaster administration mechanism has evolved over the years. Besides, a number of organisations who supplement the efforts of the government at Central, State and District levels provide vital input during emergencies and for preparedness and rehabilitation measures have also been now being institutionalised. This article provides an overall view of the disaster administration mechanism in the country at the Central, the State and the District levels also highlighting the role played by the secondary institutions. There were significant development took place in this area since last two decades, which are also highlighted in the paper.

Introduction

The unique geo-climatic conditions of the Indian sub-continent make this region among the most vulnerable to natural disasters in the world. Disasters occur with amazing frequency and while the community at large has adapted itself to these regular occurrences, the economic and social costs continue to mount year after year. According to World Bank estimates India loses about 2 of its GDP and 12 per cent of its revenues every year due to losses from natural disasters.

Indeed, concurrent to these occurrences, the governments at various levels too, have responded by taking appropriate measures for prevention and mitigation of the effects of natural disasters. While long-term preventive and preparedness measures have been taken up, the unprecedented nature of the disasters has called in for a nation-wide response mechanism wherein there is a pre-set assignment of roles and functions to various institutions at Central, State and the District levels.

India has a parliamentary democracy with a federal structure. An integrated disaster management mechanism exists within this government framework. The essential responsibility of disaster management lies with the State Government where the disaster has occurred, however, in event of disasters which are spread over several states and with uncontrollable proportions, the Central Government may be required to supplement taking appropriate measures in rescue, relief and preparedness. The change in orientation from a relief-centred to a mitigation and prevention approach has manifested in significant changes in the policy and operational levels in the recent years. Till June 2002, the Ministry of Agriculture was the nodal ministry for dealing with all natural disasters. With the shifting of the subject of disaster management to the Ministry of Home Affairs in 2002, a change in orientation has been brought about with emphasis on disaster prevention, mitigation and preparedness.

The Indian subcontinent is highly vulnerable to Drought, Floods, Cyclones and Earthquakes, though Landslides, Avalanche and Forest fires too frequently occur in the Himalayan region of northern India. Added to these are the man-made disasters like industrial accidents, road, rail and air accidents, epidemics and pandemics, civil strife etc. Among the 35 States/Union Territories in the country, 25 are disaster-prone. The States are further sub-divided into administrative units called districts—there are a total of 602 districts, of which 271 are most disaster-prone.

Floods

Of the total annual rainfall in the country, 75 per cent is concentrated over a short monsoon season of three to four months. As a result there is a very heavy discharge from the rivers during this period causing widespread floods. About 40 million hectares are vulnerable to floods, of which as much as 6.7 million hectare of land is prone to annual flooding. . The maximum affected area by floods was 17.53 million hectare in 1978.

Drought

It is perennial feature in some states of India. Sixteen per cent of the country's total area is drought-prone and approximately 50 million people are annually affected by droughts. In fact, drought is a significant environmental problem too as it is caused by a less than average rainfall over a long period of time. In India about 68 per cent of total sown area of the country is drought-prone. Most of the drought-prone areas identified by Government of India lie in the arid, semi-arid and sub-humid areas of the country.

Cyclones

India has a very long coast line of 8041 km, a major part of which is exposed to tropical cyclones arising in the Bay of Bengal and Arabian Sea. The Indian Ocean is one of the six major cyclone-prone regions of the world. In India cyclones occur usually between April and May, and also between October and December. The eastern coast line is more prone to cyclones as about 80 per cent of total cyclones generated in the region hit there. The Orissa Supercyclone of October 1999 is considered as one of the worst, in which about 10,000 people lost their lives. In this cyclone wind speeds measuring 300 km/hr, accompanied by tidal waves over 15 meters high, moved inland up to 12 km, affecting 3.4 million inhabitants.

Earthquakes

These are considered to be one of the most dangerous and destructive natural hazards. The impact of this phenomenon is sudden with little or no warning making it just impossible to predict it or make preparations against damages and collapses of buildings and other man-made structures. About 60 per cent of total area of the country is vulnerable to seismic activity of varying intensities. Most of the vulnerable areas are generally located in Himalayan and sub-Himalayan regions, and in Andaman and Nicobar Islands. The memories of recent earthquake of Latur (September 30, 1993), Bhuj (January 26, 2001and Kashmir (8 October, 2005) are still fresh in the minds of the people for the heavy damages due to house collapses and heavy loss of human lives.

The Administrative Response Mechanism

Central

In the federal set-up of India, the responsibility to formulate the Government's response to a natural calamity is essentially that of the concerned State Government. However, the Central Government pitches in with its resources. physical and financial, to provide the needed help and assistance to buttress relief efforts in the wake of major natured disasters. The dimensions of the response at the level of Central Government are determined in accordance with the existing policy of financing the relief expenditure and keeping in view the factors like:

1. The gravity of a natural calamity.
2. The scale of the relief operation necessary, and
3. The requirements of Central assistance for augmenting the financial resources at the disposal of the State Government.

At the national level, the Ministry of Home Affairs is entrusted with the nodal responsibility of managing disasters. However, in view of the highly technical and specific nature of response of technological disaster events like aviation disasters, rail accidents, chemical disasters etc, ministries dealing with the particular subject have the nodal responsibility of handling that particular disaster.

Nodal Ministries for Managing Different Types of Disasters

Within the Ministry of Home Affairs, the Central Relief Commissioner (CRC) is the nodal officer to coordinate relief operations for natural disasters. The CRC receives information on Early Warning and forecasting from the India Meteorological Department (IMD) and Central Water Commission (CWC) on a continuing basis. Other ministries/departments/organizations concerned with primary and secondary functions relating to the management of disasters constitute the Crisis Management Group (CMG).

Type of Disaster/ Crisis	Nodal Ministry
Earthquakes and Tsunami	MHA/Ministry of Earth Sciences/India Met Dept
Floods	MHA/Ministry of Water ReS/Central Water Commission
Cyclones	MHA/Ministry of Earth Sciences/India Met. Dept.
Drought	Ministry of Agriculture
Biological Disaster	Ministry of Health and Family Welfare
Chemical Disaster	Ministry of Environment and Forests
Nuclear Disasters	Ministry of Atomic Energy
Air Accidents	Ministry of Civil Aviation
Railway Accidents	Ministry of Railways

Source: National Disaster Management Authority (www.ndma.gov.in).

A nodal officer, nominated from each ministry/department is responsible for preparing the sectoral action plan/Emergency Support Function Plan for managing disasters. The CMG's functions are to review the Contingency plans, measures required for dealing with natural disasters, coordinate the activities of the Central Ministries and State governments in relation to disaster preparedness and relief. In the event of a disaster, the CMG meets frequently to review the relief operations and extend all possible assistance required by the affected states to overcome the situation effectively. National Crisis Management Committee (NCMC): Under the chairmanship of the cabinet secretary the NCMC has been constituted in the cabinet secretariat. The other members of this committee include the Secretary to Prime Minister, Secretaries of Ministry of Home Affairs, Defence, Research and Analysis Wing and Agriculture and Cooperation along with Director Intelligence Bureau and an officer of cabinet secretariat. The NCMC gives direction to the crises management group as deemed necessary.

At the apex level, there are two Cabinet Committees *viz.* the **Cabinet Committee on Natural Calamities (originally the Cabinet Committee on Drought Management) and the Cabinet Committee on Security**. Major issues relating to natural disasters, primarily pertaining to the institutional and legislative measures for an effective strategy for natural disaster management are placed before the Cabinet Committee on Natural Calamities. In case of calamities which impinge on internal security or which may be caused due to use of nuclear, biological and chemical weapons/materials, the matter is required to be placed before the Cabinet Committee on Security.

State

At the State level, the State Relief Commissioner (or Secretary, Department of Revenue) supervises and controls relief operations through Collectors or Deputy Commissioners, who are the main functionaries to coordinate the relief operation at district level.

As pointed out earlier, the Central Government only supplements the efforts of the State Governments. The State Governments are autonomous in organising relief operations in the event of natural disaster and in the long-term preparedness/ rehabilitation measures.

The States have Relief Commissioners who are in charge of the relief measures in the wake of natural disasters in their respective States. Recently, in accordance with the directives of the Central Governments, the nomenclature of the Relief Departments in many states has been changes to Disaster Management Departments. In the absence of the Relief Commissioner, the Chief Secretary or an Officer nominated by him is in overall charge of the Relief operations in the concerned State.

The Chief Secretary is the head of the State Administration. The State Headquarters has, L.'1 additio:1, a number of secretaries, heads of the various Departments handling specific subjects under the overall supervision and coordination of the Chief Secretary. While important policy decisions are taken. at the State Headquarters by the Cabinet of the State headed by the Chief Minister. Day-today decisions involving policy matters are taken or exercised by the Secretary in the Department.

States Crisis Management Group: There is a State Crisis Management Group (SCMG) under the Chairmanship of Chief Secretary/Relief Commissioner . This group comprises Senior Officers from the Departments of Revenue/Relief, Home. Civil Supplies, Power, Irrigation, Water Supply, Panchayat (local self-Government), Agriculture, Forests, Rural Development, Health Planning, Public Works and Finance.

The SCMG is required to take into consideration the infrastructure and guidance received, from time to time, from Government of India and formulates action plans for dealing with different natural disasters.

It is also the duty of the Relief Commission of the State to establish an emergency operation Centre as soon as a disaster situation develops. Besides having all updated information on forecasting and warning of disaster the Centre would also be the contact point for the various concerned agencies.

District

States are further divided into districts, each headed by the District Collector (also known as the District Magistrate or Deputy Commissioner). It is the District Collector who is the focal point at the District level for directing, supervising and monitoring relief measures for disaster and for preparation of District level plans.

The Collector exercises coordinating and supervisory powers over functionaries of all the Departments at the District level. During actual operations for disaster mitigation or relief, the powers of the Collector are considerably enhanced, generally, by standing instructions or orders on the subject, or by specific Governments orders, if so required. Sometimes, the administrative culture of the State concerned permits, although informally, the collector to exercise higher powers in emergency situations and the decisions are later ratified by the competent authority.

A District is sub-divided into sub-divisions and *Tehsils* or *Talukas.* The head of a subdivision is called the Sub-Division Officer (SDO) while the head of a *Tehsil* is generally known as the *Tehsildar* (*Talukdar* or *Mamlatdar* in some States). Contact with the individual villages is through the Village Officer or *Patwari* who has one or more villages in his charge When a disaster is apprehended, the entire machinery of the District, including officers of technical and other departments, swings into action and mailli"ins almost continuous contact with each village in the disaster threatened area. In the case of extensive disasters like drought, contact is maintained over a short cycle of a few days.

The various measures undertaken by the District Administration area are as follows:

Contingency Plans

At the district level, the disaster relief plans are prepared which provide for specific tasks and agencies for their implementation in respect of areas in relation to different types of disasters.

A contingency plan for the district for different disasters is drawn up by the Collector Deputy Commissioner and approved by the State Government. The Collector / Deputy Commissioner also coordinates and secures the input from the local defence forces unit in preparation of the contingency plans. These contingency plans lay down specific action points, key personnel and contact points in relation to all aspects.

District Relief Committee

The relief measures are reviewed by the district level relief committee consisting of official and non-official members including the local legislators and the members of parliament.

District Control Room

In the wake of natural disasters, a Control Room is set up in the district for day to day monitoring of the rescue and relief operations on a continuing basis.

Coordination

The Collector maintains close liaison with the Central Government authorities in the districts, namely, the Army, Air Force and Navy, Ministry of Water Resources, etc., who supplement the effort of the district administration in the rescue and relief operation.

The Collector Deputy Commissioner all coordinates voluntary efforts by mobilizing the non-governmental organizations capable of working in such situations.

The entire hierarchy right from the Central Government (the Department of Agriculture and Co-operation in the Ministry of Agriculture) to the District level, and even the sub *Divisional Tehsil level*, is connected with a telecommunication system. The normal mode of telecommunications is overland telephone and telegraphy, but in times of stress and if there is a breakdown of the overland system, radio communication is resorted to. The wireless network is generally run and maintained by the police organisation in the country.

The Recent Developments

The Disaster Management Act 2005 was one of the most significant initiatives taken by the Government of India for putting in place an institutional system dedicated to disaster management. Notified on 26 December, 2005, the Act was the first acknowledgement of the Government of India of the need for legislative backup to the governance system. Comprising of 79 sections and 11 chapters, the Act provides for the "effective management of disasters and for matters connected therewith or incidental thereto." The Disaster Management Act 2005 has created a hierarchy of institutions at the national, state and district levels for holistic management of disasters. The national level organizations created as per the Act are:

National Disaster Management Authority (NDMA)

With the Prime Minister as Chairperson and 9 other members the National Authority (NDMA) is entrusted with the responsibility of laying down the policies, plans and guidelines for ensuring timely and effective response to disasters. More specifically, the National Authority is mandated to lay down policies, approve the national plan, approve plans of other ministries/departments, lay down guidelines for the states, coordinate implementation of policies and plans, recommend mitigation mitigation funding provisions and coordinate bilateral support to other affected countries during disasters. The National Authority is also expected to frame guidelines for provision of minimum standards of relief, special provisions to be extended to widows and orphans and ex gratia assistance for restoration.

Similarly, every state is supposed to have State Disaster Management Authority (SDMA), headed by the Chief Minister of the state and every District is to have District Disaster Management Authority (DDMA) headed by the District Magistrate and President of Zila Parishad (Public Representative).

National Executive Committee (NEC)

The Act provides for the constitution of a National Executive Committee under the chairmanship of the Home Secretary to assist the Authority in performance of its functions. The National Executive Committee would comprise of the Secretaries to the ministries/departments of agriculture, defence, drinking water supply, environment and forests, finance (expenditure), health, power, rural development, telecommunication, space, science and technology, urban development, water resources and the Chief of the Integrated Defence Staff of the Chiefs of Staff Committee.

National Institute of Disaster Management (NIDM)

NIDM was founded from its predecessor National Centre for Disaster Management (NCDM) with an aim of creating an Institute of excellence in disaster management studies in India. NIDM is also required to network with various research and training institutions for sharing of knowledge and resources.

National Disaster Response Force

The National Disaster Response Force (NDRF) has been constituted by upgradation/ conversion of 8 standard battalions of the Central Para Military Forces as specialist force to respond to disaster situations. The NDRF has been carved out 2 battalions each from Border Security Force (BSF), Indo-Tibetan Border Police (ITBP), Central Industrial Security Force (CISF) and Central Reserve Police Force (CRPF). Based on the vulnerability profile of different regions of the country, these battalions are stationed at 9 different locations to be deployed, in the event of any serious threatening disaster situation, to provide instantaneous response.

A similar institutional structure has been created at the State level with each state to have State Disaster Management Authorities under the chairpersonship of the Chief Minister, State Executive Committee under the chairpersonship of the Chief Secretary and four secretaries of relevant departments, State Disaster Response Forces for disaster response. The existing resources of the state armed police, fire and rescue services, Home Guards, Civil Defence, etc., would be the sources from which the SDRFs may be constituted to generate specialist response. They will also include women members for looking after the needs of women and children.

At the district level, District Disaster Management Authorities (DDMA) under the co-chairpersonship of the District Magistrate and the President of the Zila Parishad, will provide for integration of the executive and legislative focal points at the district level. Their major responsibilities would include:

1. Preparation of district disaster management plan including district response plan,
2. Coordination and monitoring implementation of the national and state policies and plans,
3. To take requisite measures for disaster prevention and mitigation in the vulnerable areas,
4. To give directions to concerned departments for putting in place risk reduction measures,
5. To organize capacity building of the staff,
6. To facilitate community training and awareness,
7. To coordinate early warning and dissemination mechanisms,
8. To establish stockpiles of relief and rescue materials,
9. To ensure regular rehearsals, drills etc., and
10. Communicating with their State Authority.

The Act has also provided for the setting up of different funds at the national, state and district levels At the national level, **The National Disaster Response Fund**, as specified in the Act, would be used for "meeting any threatening disaster situation or disaster" while **The National Disaster Mitigation Fund** has been provided *exclusively for the purpose of mitigation* and would be used only for mitigation projects. The corpus of the fund would be provided by the Central Government after due appropriation made by Parliament, by law.

State Funds

Similar funds are to be provided at the state and district levels. The Act enjoins the State Governments to create Response and Mitigation Funds at the state and district levels. **The State Disaster Response Fund** would available for emergency response, relief and rehabilitation at the state level, while **The State Disaster**

Mitigation Fund, to be made available to the State Disaster Management Authority for mitigation projects.

District Funds

At the district level, the State Government has to create similar funds for the district level. **The District Disaster Response Fund** and **The District Disaster Mitigation Fund** would be made available to the District Authorities for response and mitigation purposes respectively.

Besides the district officials, a host of other bodies too supplement their efforts in disaster situations–particularly the armed forces and the non-government voluntary organizations as follows:

The Armed Forces

The armed forces of the country have played a vital role during disaster emergencies providing prompt relief to the victims even in the most inaccessible and remote areas of the country. The organisational strength of the armed forces with their disciplined and systemized approach, and with their skills in technical and human resource management make them indispensable for such emergency situations.

Besides when disasters occurs over large areas, it is usually beyond the capability of the administration to organise the relief activities, the armed forces are then called upon to organise the relief measures.

Related to the efforts of the armed forces, are the Civil Defence and the Home Guard Organisations. These organisations are voluntary in nature and character and come in handy in emergency situations like natural disasters. A network of these is now found all over the country . Their aim, while not actually taking part in actual combat operations like in army, is firstly to save lives, to minimise damage to property and to maintain continuity to production. Thus, while disaster situations often lead to chaotic conditions where rescue and relief work is severely affected. These organisations are able to co-ordinate and support efforts in a disciplined manner so that both the army and the district officials are able to carry out their respective activities efficiently.

Non-Government Voluntary Organizations

Emerging trends in managing natural disasters have highlighted the role of Non-Governmental Organisations (NGOs) as one of the most effective *alternative* means of achieving an efficient communication link between the Disaster Management agencies and the affected community. Many different types of NGOs are already working at advocacy level as well as grassroots level; in typical disaster situations they can be of help in preparedness, relief and rescue, rehabilitation and reconstruction and also in monitoring and feedback.

The role of NGOs is a potential key element in disaster management. The Non-Governmental sector that operate at grassroots level, can provide a suitable alternative as they have an edge over Governmental agencies for invoking community involvement. This is chiefly because, the NGO sector has strong linkages with the community base, and can exhibit great flexibility in procedural matters vis-a-vis the government.

Based on the identified types of NGOs and their capabilities, organised action of NGOs can be very useful in the following activities in different stages of disaster management:

1. Awareness and information campaigns,
2. Training of local volunteers,
3. Advocacy and planning,
4. Immediate rescue and first-aid including psychological aid,
5. Supply of food, water, medicines, and other immediate need,
6. Materials,
7. Ensuring sanitation and hygiene,
8. Damage assessment,
9. Technical and material aid in reconstruction Assistance in seeking financial aid, and
10. Monitoring.

Research and Training Institutions

In India, a number of Research Institutes are conducting active research in the field of Disaster Management. Valuable inputs in technical, social, economic as well as management areas of the field are being looked into.

Research activities are being coordinated by different ministries depending on the type and level of research. An important role is played by the Universities too in this sector who, besides running programmes on disaster management, also serve as think tanks for the government. Indira Gandhi National Open University (IGNOU) conducts a certificate course on disaster management since 1998. Institutes spread geographically across the country have developed specialisation in terms of particular regions where most of their research is concentrated and also in terms of particular disasters. Notable Universities are the University of Roorkee, The Indian Institutes of Technology Delhi and Chennai and the Anna University, Chennai.

The Department of Science (Ministry of Science and Technology), Government of India coordinates activities through a network of scientific institutes, *e.g.*, the Central Building Research Institute at Roorkee. The Ministry of Urban Development carries out research through the Building Materials and Technology Promotion Council, on subjects such as appropriate building materials for disaster-prone areas. These institutes, besides providing technical assistance to implementing and engineering organisations also train field level officers/and other concerned role players.

The Community

Recent trends have revealed that the community as an institution in itself is emerging as the most powerful in the entire mechanism of disaster administration. In event of actual disasters, the community, if well aware of the preventive actions it is required to take, can substantially reduce the damage caused by the disaster.

Awareness and training of the Community is particularly useful in areas which are prone to frequent disasters.

It is laudable, the efforts in certain areas where communities have formed their own organisations which are take the right initiative in such situation. One such community based organisation is the Village Task Force formed in villages of Andhra Pradesh by the Church Auxiliary for Social Action (CASA). The Village Task Force has been trained in emergency evacuation and relief within the village. It is elected by the people themselves and during disasters it serves as the nodal body at village level which has to mobilise resources for the community and disseminate necessary information passed on by the outside agencies. The UNDP led DRM programme along with Govt. of India covers 160 multi-hazard prone districts has a focus towards community based disaster risk management programme.

While the community as an effective institution is yet to take shape in this country with low literacy levels and widespread poverty, considerable efforts are being made to form and strengthen community based organisations at grass root levels.

Conclusion

In disaster situations, a quick rescue and relief mission is inevitable, however considerable damage can be minimised if adequate preparedness levels are achieved. Indeed, it has been noticed in the past that as and when attention has been given to adequate preparedness measures, the loss to life and property has considerably reduced. Preparedness measures such as training of role players including the community, development of advanced forecasting systems, effective communication and above all a well-networked institutional structure involving the government organisations; academic and research institutions, the armed forces and the NGOs would greatly contribute to the overall disaster management of the region. The government's recent policy changes too reflect the changing approach from *rescue and relief to preparedness*.

Chapter 5

Policy Issues and Mitigation Strategies of Natural Disasters in Malaysia

Puan Hajah Che Gayah Binti Ismail
Malaysian Meteorological Department,
46667 Petaling Jaya, Selangor Darul Ehsan, Malaysia
E-mail: *cgayah@kjc.gov.my*

ABSTRACT

The author here discusses the efforts being made by the Malaysian Government to disaster risk reduction possibilities by development of disaster management capabilities in the various areas of disaster prevention, mitigation, response and recovery as well as having a systematic development and application of policies, strategies and practices to minimize vulnerabilities, hazards and impact of disaster on a society in the broad context of sustainable development. The prime objective of the paper is to highlight some strategies on the natural disaster risk reduction possibilities.

Malaysia has undertaken steps toward establishing National Disaster Database Information Centre in Malaysian Centre for Remote Sensing (MACRES). It can be concluded that the Government must also enhance its national capabilities in disaster management by the setting up of training facilities and through related activities in cooperation with other countries.

Introduction

A natural disaster is a serious disruption to a community or region caused by the impact of a naturally occurring rapid onset of an event that threatens or causes death, injury or damage to property or the environment and which requires significant and coordinated multi agency and community response. Although Malaysia is geographically located outside the "Pacific Rim of Fire" and is relatively free from any severe ravages destruction caused by natural disasters such as earthquake, typhoons and volcanic eruptions, nevertheless the country has experienced other types of natural disasters such as the monsoon floods, landslides and severe haze. Malaysia also went through several extreme climatic events, ranging from freaky thunderstorms to monsoonal floods.

Malaysian Government is very concerned about occurrences of such disasters that adversely affect her people. Emphasis is being given to disaster risk reduction possibilities by development of disaster management capabilities in the various areas of disaster prevention, mitigation, response and recovery as well as having a systematic development and application of policies, strategies and practices to minimize vulnerabilities, hazards and impact of disaster on a society in the broad context of sustainable development.

Objective

In addition to giving brief outline and highlights on the machinery and mechanism of the current disaster management in Malaysia, the main objective of this paper is to highlight some strategies on the natural disaster risk reduction possibilities.

Challenges Posed by Disasters

Disaster loss is on the rise with grave consequences for the survival, dignity and livelihood of individuals, particularly the poor and hard-won development gain.

Disaster risk and vulnerabilities are on the rise due to changing demographic, technological and socio-economic conditions, unplanned urbanization, development within high-risk zones, under development, environmental degradation, climate variability, climate change and geological hazards.

There is now international acknowledgement that efforts to reduce disaster risks must be systematically integrated into policies, plans and programs for sustainable development and poverty reduction and supported through bilateral, regional and internationals cooperation, including partnerships. The importance of promoting disaster risk reduction efforts on the international and regional levels as well as the national and local levels has been recognized in the past few years in a number of key multilateral frameworks and declarations.

Studies have shown that developing countries suffer far greater economic lost because of natural disasters as compared to developed countries. World Bank estimates that the annual cost of natural disasters to developing countries to be in the region of 2 to 15 per cent of GDP. Study by the United State Geological Survey and the World Bank estimated an investment of US40billion in disaster risk reduction would prevent losses of US280 Billion in the 1990s. World Meteorological Organization finding showed that during the period 1992-2001, natural disasters world wide killed over 622,000 and affected more than two billion people. The WMO also estimated that a dollar spent on disaster preparedness can prevent seven dollars of disasters-related losses.

General literature on natural disasters also indicated that there are significant benefits to be realized from well developed Disaster Risk Reduction (DRR) strategies.

Average fatalities caused by major earthquakes in developed countries have fallen from about 12,000 in the period 1900 to 1949 to 2,000 in 1950 to 1992 largely as a result of better structural engineering and preparedness. However, the average remained constant at 12,000 in developing countries. Studies have also shown that natural disasters have major impacts at the macro level such as value of lost output, diverting government funds for disaster relief efforts and increased indebtedness and also demonstrate the potential for economically effective DRR measures in developing countries.

These and many other findings of research done on DRR strategies in developing countries prone to natural disasters indicated huge benefit can be derived from a well develop DRR strategies. As a developing country, Malaysia is also facing a higher risk of losses and damages due to natural disasters. Floods are the primary hazards faced by Malaysia whereby about 12 per cent of the area in Peninsular Malaysia, where almost 2.5 million people live in the western low plains of muddy settlements, is flood prone. Along the sandy beaches and bays of the east coast of Peninsular Malaysia, more than 200,000 people or another 12,600 square kilometre are under flooding risk. The causes are a combination of excessive precipitation and runoff, inadequate river capacity, flood plain encroachments, blocking of river mouths and high tides. It is estimated that river flooding, heightened by tides, causes an estimated annual damage of RM38 million and the latest Tsunami had caused loses and damages of more than RM7.5 million. Table 5.1 is the summary of disasters affecting Malaysia during 1965 – 2004 periods. The sources of the data are the Centre for Hazards and Risk Research at Columbia University (Cyclone, Drought and Flood) and The National Security Council Malaysia (tsunami).

Table 5.1: Summary of Disasters Affecting Malaysia during 1965–2004.

Disaster	No. of Events	Total Killed	Total People Affected
Cyclone	6	294	55,805
Drought	1	0	5,000
Flood	24	243	899,620
Tsunami	1	68	7,721

Priorities for Action

The boundary between natural and man made disaster is often blurred as natural disaster can be caused by natural hazard or a combination of both, for example land slides which are caused by a combination of deforestation, heavy rain and light earth tremor. Thus in general the strategies proposed will cover both natural and man-made disasters.

Role of Malaysia National Security Department

In Malaysia National Security Department (NSD) is responsible for coordinating activities related to the preparation, prevention, response and handling of all type of disasters and using the National Security Council (NSC) Directive No. 20 on "Policy and Mechanism on Natural Disaster and Relief Management" as guidelines. Directive

No. 20 gives outlines on Disaster and Relief Management according to the level and complexity of disaster, and to establish the management mechanism with the purpose of determining the roles and responsibilities of the various agencies involved in handling the disaster.

The handling and resolving of disasters in Malaysia are currently through the Committee System which emphasizes on the concept of coordination and mobilizations of agencies involved in an integrated and coordinated manner. In this respect, the management and handling of disaster are undertaken by the Disaster Management and Relief Committee at the Federal, State and District levels respectively. Malaysian Meteorological Department is a member of the committee at all levels and is responsible in providing meteorological related information and advises. The Committee as stipulated in the NSC Directive No. 20 ensures that the threat to public safety as well as protection of properties are effectively managed and handled.

Existing Policies

In order to minimize the impact of development to the environment which in turn will expose the population or place to a higher risk of natural disasters the government has created acts and guidelines on development such as The Environment Act of 1987 which requires all large scale development projects to prepare Environment Impact Assessments prior to project approval and the Fisheries Act of 1985, South Johore Coastal Resources Management Plan and Integrated Coastal Zone Management for coastal development and industries.

Even though the mechanisms in preparation, prevention, response and mitigating disasters are in place however there seems to be something amiss in the way disaster such as flash flood keep on occurring within the same area and the inability of the authority to handle the disaster for example the Shah Alam flash flood on the 26th February 2006 which affected more than 4000 houses and caused damages of more than five millions (RM). According to a report established by Department of Irrigation and Drainage at the meeting in the State Secretary Office on the 28th February, 2006, even though the flood warning siren for the Shah Alam flash flood was activated, the affected community were unaware of it.

Flash floods also seem to be the norm for Kuala Lumpur city whenever there is a heavy downpour of several hours. Thus there should be a holistic approach towards disaster risk reduction measures in all development and maintenance programmes. The approach should not only cover the aspect of disaster management as of Directive No. 20 but also on community participation such as public education and awareness programme.

Future Programs for Disaster Risk Reduction

In formulating national guidelines and policies, Malaysia should learn from the experiences of developed countries and use their guidelines and policies as references. We should also look at guidelines, frameworks and policies proposed and adopted at the International level such as Hyogo Framework for Action 2005-2015: Building the Resilience of Nations and Communities to Disasters and a policy paper of Department of International Development (DFID) on Reducing the Risk of Disasters – Helping to Achieve Sustainable Poverty Reduction in a Vulnerable World.

In order to enhance Malaysia's capability in disaster risk reduction in the future the country should adopt the following strategies:

1. Ensuring that disaster risk reduction is a national and a local priority with a strong institutional basis for implementation;
2. Identifying, assessing and monitoring risks and enhance early warning;
3. Using knowledge, innovation and education to build a culture of safety and resilience at all levels;
4. Reducing the underlying risk factors;
5. Strengthening disaster preparedness for effective response at all level.

Countries that develop policy, legislative and institutional framework for disaster risk reduction and that are able to develop and track progress through specific and measurable indicators have greater capacity to manage risks and to achieve widespread consensus for, engagement in and compliance with disaster risk reduction measures across all sectors of society. Key activities that should be carried out include having national institutional and legislative framework, resources management and community participation.

The legislative framework should support the creation and strengthening of national integrated disaster risk reduction mechanisms, such as multi sectoral national platforms with designated responsibilities at the national through the local levels to facilitate coordinating across sectors.

National Security Division as the coordinating agency should call for a revision of the National Security Council Directive No. 20 as to ensure that the Directive is still relevant and covers all type of disasters including tsunami related disasters.

The starting point for reducing disaster risk and for promoting a culture of disaster resilience lies in the knowledge of the hazards and the physical, social, economic and environmental vulnerabilities to disasters that most societies face, and of the ways in which hazards and vulnerabilities are changing in the short and long term, followed by action taken on the basis of that knowledge. The key activities should include national and local risk assessments, early warning, capacity development and regional and emerging risks.

National and local risk assessment includes development, updating and dissemination of risk maps and other related information to decision makers, general public and communities at risk in an appropriate format. Early warning systems must be people cantered, in particular system whose warnings are timely and understood to those at risk, which take into account the demographic, gender, cultural and livelihood characteristics of the target audiences including guidance on how to act upon warnings and that support effective operations by disaster managers and other decision makers.

As the most severe climate related natural disasters in Malaysia are monsoon floods and flash floods, the government under DID with the assistance of Canadian Government has established the Flood Forecasting and Warning System. The system comprises of 233 telemetric rainfall stations, 190 telemetric water level stations, 256 manual stick gauges, 84 flood warning boards, 217 flood sirens, Real-time flood forecasting and warning systems in nine river basins. The system, known as Infobanjir is web-based and the information is available to the general public who has access to the Internet.

Malaysian Meteorological Department is in the process of completing the Tsunami Early Warning System project which was approved after the 2004 Tsunami disaster. The system will enable the potential affected people to receive warnings by various means such as siren, cell and fixed line broadcast and media. With additional allocation, the system can be expanded to cover other disasters as well.

Capacity development includes support for development and sustainability of the infrastructure and databases, improvement of scientific, technical methods and capacities for risk assessment, monitoring and early warning and establishment and strengthening of the capacity for data analyzing and dissemination. There should also be capacity development in research, analysis, report and support for development of common methodologies for risk assessment and monitoring.

Disasters can be substantially reduced if people are well informed and motivated towards a culture of disaster prevention and resilience, which in turn requires the collection, compilation and dissemination of relevant knowledge and information on hazards, vulnerabilities and capacities. Some of key activities that should be undertaken are information management and exchange, education and training, research and public awareness.

Malaysia has undertaken steps toward establishing National Disaster Database Information Centre in Malaysian Centre for Remote Sensing (MACRES). Education, training, research and public awareness program are ongoing activities and are undertaken by relevant departments, such as Malaysian Meteorological Department for weather and seismic activities and National Security Department for relief activities and evacuation. Awareness programs that are being undertaken include lectures and briefings for school children and public, open day programs and publishing and airing of related information through the mass media.

Disaster risks related to changing social, economic, environmental conditions and land use, and the impact hazards associated with geological events, weather, water, climate variability and climate change are addressed in sector development planning and programs as well as in post-disaster situation.

Key activities to be carried out include environmental and natural resource management, social and economic development practices, land use planning and other technical measures.

Malaysia has actively promote the reduction of underlying risk factors. Some of activities being undertaken are creation of Environmental Quality Act, Publishing of Handbook of Environmental Impact Assessment Guidelines, Legislation in Planning and Land Use, Agriculture, Mining, Marine Pollution, Wildlife and Protected Areas, Fisheries, Forestry, Rights of Indigenous People, coastal zone management and strengthened enforcement of regulation for development. Projects include building of retention ponds and raising road levels in flood prone areas in Shah Alam, smart tunnel for Kuala Lumpur and national coastal study.

Malaysia should also promote the integration of risk reduction associated with existing climate variability and future climate change into strategies for the reduction of disaster risk and adaptation to climate change, which would include the clear identification of climate related disaster risks, the design of specific risk reduction measures and an improved and routine use of climate risk information by planners, engineers and other decision makers.

At times of disaster, impact and losses can be substantially reduced if authorities, individuals and communities in hazard prone areas are well prepared and ready to act and are equipped with knowledge and capacities for effective disaster management.

Regional and sub-regional cooperation in the field of disaster management and preparedness is recognized as very important in the effort to reduce disaster risk and to implement effective mitigation, response and recovery measures. In this respect, Malaysia gives support to the various regional forums such as the ASEAN Committee on Disaster Management (ACDM), the ASEAN Regional Forum (ARF), the ASEAN Regional Haze Task Force as well as the Sub-Regional Fire Fighting Arrangement. Malaysia believes that regional mechanism is needed to promote exchange of information and sharing of experiences among personnel involved in disaster management in the region. Such regional cooperation is expected to contribute greatly to disaster reduction efforts and enhance national disaster management capabilities in the areas of resource mobilization, communication networks, warning system, forecasting techniques and training.

Conclusion

Due to the rapid development in the country, the cost of damage due to disaster is also escalating. Malaysia must realize the importance that the element of safety must be built into each development project to avoid high casualty and damage to life and property in the event of a disaster. An integrated approach in the various aspects of disaster management involving the various agencies must be adopted to achieve higher efficiency and greater effectiveness.

Efforts must be made by the Government to create higher awareness among the people so as to enhance disaster preparedness as well as to build a culture of prevention and Civil Protection/Public Safety in the community. The Government must also enhance its national capabilities in disaster management by setting up of training facilities and through related activities in cooperation with other countries. To be a developed country, Malaysia has to ensure that all the development efforts not gone to waste due to natural disasters. Malaysia must create a safer environment for the people through effective management, risk reduction efforts and sustainable development in the 21st century.

References

[1] Asian Disaster Preparedness Centre http://www.adpc.net/oed/Country per cent 20 Profiles/ Malaysia.pdf.

[2] Banjir kilat di Shah Alam, http://ms.wikipedia.org/eiki/ Banjir_kilat_di_Shah_Alam.

[3]   DRI- Disasters Reduction Information Analysis Tool http://gridca.grid.unep.ch/undp/.

[4]   Environmental Management and the Mitigation of Natural Disasters, http://www.un.org/womenwatch/daw/csw/env_manage/documents/EGM-Turkey-final-report.pdf.

[5]   Flood relief centres in Perlis, Terengganu close, http://www.nst.com.my/Current_News/NST/Wednesday/National/20051228093658/Article/pp_index_html.

[6]   Guide On Improving Public Understanding Of And Response To Mitigation and Management of Flood Disasters in Malaysia, http://www.adrc.or.jp/publications/TDRM2005/TDRM_Good_Practices/PDF/Chapter3_3.3.6.pdf.

[7]   Hyogo Framework for Action 2005-2015, http://www.unisdr.org/wcdr/intergover/ official-doc/L-docs/Hyogo-framework-for-action-english.pdf.

[8]   International Day For Natural Disaster Reduction, http://www.unescap.org/esd/water/ events/idndr/2002.

[9]   Natural Disaster Mitigation Program, www.disaster.qld.gov.au/disasters/mitigation.asp.

[10] Natural Disaster and Disaster Risk Reduction Measures, http://www.unisdr.org/news/ DFID-Economics-Study-for-DfID.pdf.

[11] Natural Disaster Prevention and Mitigation (DPM), http://www.wmo.ch/disasters/.

[12] The Malaysian Experience and Future Direction on Disaster Management, http://www.adrc.or.jp/countryreport/MYS/2003/page2.html.

[13] Response to Dynamic Flood Hazard Factors in Peninsular Malaysia, http://www.questia.com/PM.qst?a=o&se=gglsc&d=5001641532

[14] Reducing the Risk of Disaster – Helping to Achieve Sustainable Poverty Reduction in a Vulnerable World : A DFID policy paper, http://www.dfid.gov.uk/pubs/files/ disaster-risk-reduction-policy.pdf

[15] Reducing Vulnerability to Weather and Climate Extremes, http://www.unisdr.org/eng /media-room/ statements/stmts-2002-22-march-USG.doc.

[16] Shah Alam masih gagal atasi permasalahan banjir http://www.viewmalaysia.com.

[17] U.S Participation In International Decade For Natural Disaster Reduction, link.aip.org/link/?NHREFO/1/2/.

[18] WMO Calls for Investment in Early Warning Systems to Reduce the Toll of Natural Disasters, http://www.meteo.bg/reports/reports/wmo/PR.720_E.pdf.

[19] WMO's Roles in Disasters Mitigation and Response Challenges and Opportunities, http://www.nws.noaa.gov/iao/AMS/2004/Presentations/ Michel per cent 20 Jarraud per cent 20 Presentation.ppt#4.

Chapter 6

Country Report on Earthquake Hazard Assessment of Syria

Mohamad Daoud
Higher Institute of Earthquake Studies and Research,
Damascus University, Damascus, Syria
E-mail: *m-daoud@scs-net.org*

ABSTRACT

Points out that a number of Syrian institutions are working to assess the earthquake hazards in Syria. The National Earthquake Centre (NEC) operates a regional network since 1995. In addition, many relevant institutions such as the Atomic Energy Commission of Syria (AECS), Universities of Damascus and Tishrin, and the Higher Institute for Applied Sciences and Technology (HIAST) with cooperation of Cornell and Strasbourg Universities have been doing investigations in the field of palaeoseismology. The author here presents the summary and results of these investigations. Concludes that these multi researches should certainly contribute in mitigation of the hazards in Syria and the surrounding regions to a large extent.

Introduction

Assessment of earthquake hazards represents one of the main first steps toward developing the earthquake reduction management strategy. In order to assess the earthquake hazards in Syria, many Syrian institutions are working in this regard. The National Earthquake Centre (NEC) operates a regional network since 1995. In addition, many relevant institutions such as the Atomic Energy Commission of Syria (AECS), Universities of Damascus and Tishrin, and the Higher Institute for Applied Science and Technology (HIAST) with co-operation of Cornell and Strasbourg Universities have been doing investigations in the field of palaeoseismology. Also, AECS compiled in 1998 the historical and instrumental earthquake catalogues. All these scientific activities are for earthquake assessment and evaluation for the country. A summary of these investigations and results will be presented in paragraph 3.

General Tectonic Setting

Syria is located at the northern part of the Arabian Plate. This plate is bounded from the west by the general N-S direction Dead Sea Fault System (DSFS) with 1000 km long from the Gulf of Aqaba in the south to near Antioch in the north, where it intersects the Eastern Anatolian Fault System (EAFS). The DSFS in Syria consists of many faulting segments such as Serghaya, Missyaf and Al-Ghab faults. Between Damascus and the Euphrates River, the northern trending Palmyra fold thrust is located within the Arabian Plate. This belt consists of many asymmetrical elongated anticlines separated by narrow depressions. Figure 6.1 shows the major fault zones of the northern Arabian plate.

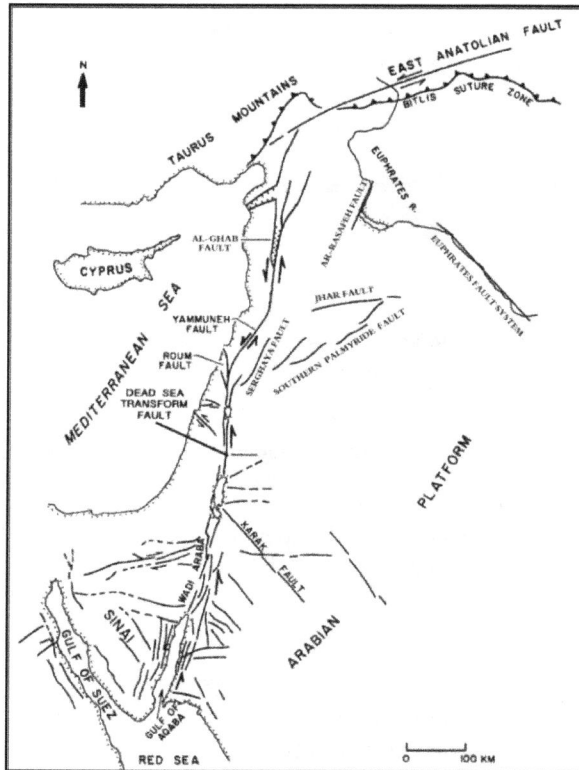

Figure 6.1: Map of the Northern Part of the Arabian Plate Showing the Major Fault Zones.

Earthquake Hazards Assessment Studies in Syria

The Syrian National Seismological Network (SNSN)

The SNSN consists of two networks: 1) 27 short-period stations, and 20 three-component accelerographs stations. They cover Syria especially its western part.

Both networks are equipped with suitable hardware and software for digital data acquisition and processing. Figure 6.2 shows distribution of 27 SP stations. They are capable to monitor the earthquake activities in and around Syria. It is worth mentioning that there is a monthly seismological bulletin issued by the NEC. In fact, the seismicity of Syria for the period 1995-2005 is moderate (M ≤ 5°). [For more information about the National Earthquake Centre and its bulletins, visit the website: www.nec.gov.sy]. Figure 6.3 shows the earthquake activity in and around Syria for the period 1995-2005.

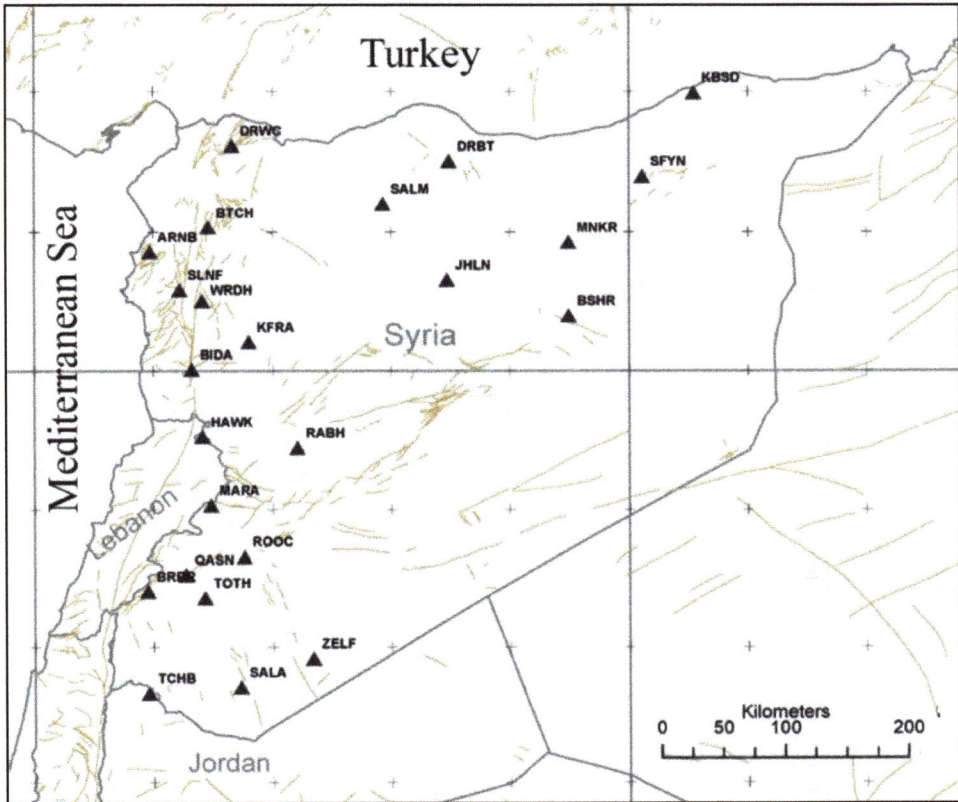

Figure 6.2: Map of Syria Showing Distribution of 27 Stations Network.

Palaeoseismological Studies

Studying the palaeoseismology along two faulting segments (namely Serghaya and Missayaf) of the northern part of the DSFS in Syria documents palaeoseismic expressions of Holocene movements on these two segments. Trenches were excavated across these faults and charcoal samples were obtained for dating the events. Results show that these two segments have ruptured during the Holocene. They are active ones and capable of generating large magnitude earthquakes of the order ≥ 7°.

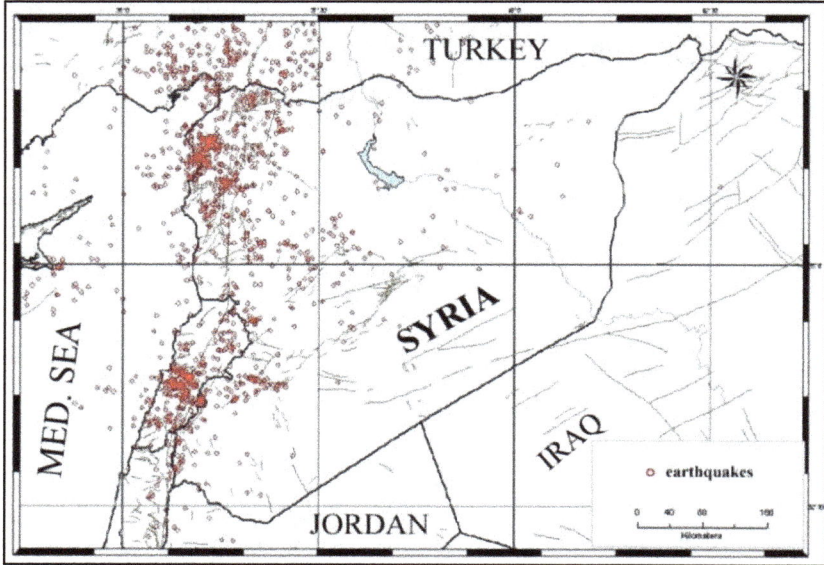

Figure 6.3: Map of Syria Showing the Earthquake Activity Recorded by NEC.

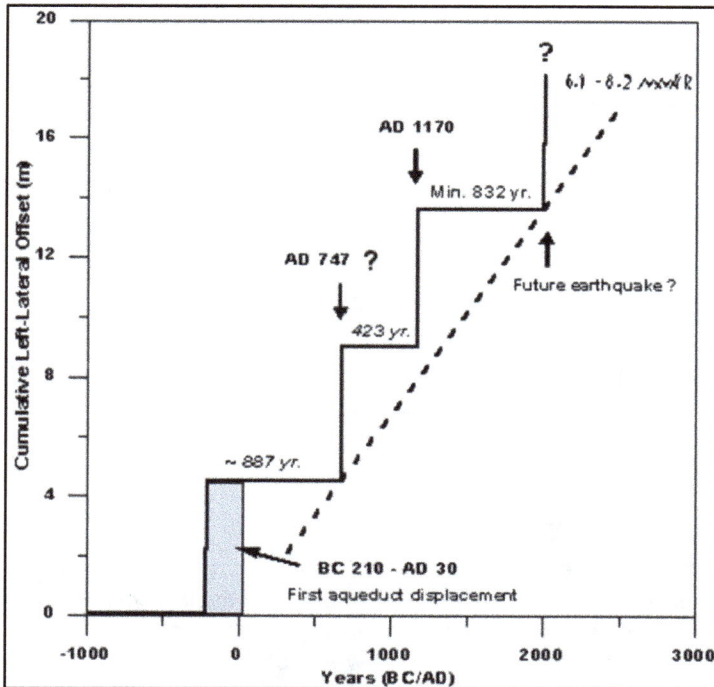

Figure 6.4: Cumulative Left-Lateral Offset Over the Last 2000 yr Along the Missyaf Segment Obtained from Palaeoseismic Study and Historical Data.

Historical Seismicity Studies

Investigating the historical earthquakes that occurred in Syria using historical documents and other sources resulted in an earthquake catalogue including 181 events for the period 1365 BC-1900 AD. The most large earthquakes occurred along the northern segments of the DSFS are in 551 AD, 1157, 1170, 1202, 1408, 1759 and 1822. These earthquakes were associated with geological hazards such as foreshocks, aftershocks, ruptures, landslides and tsunamis. Also these events caused large losses of lives in cities distributed near the DSFS in Syria and Lebanon. Figure 6.5 reveals distribution of historical seismicity in and around Syria.

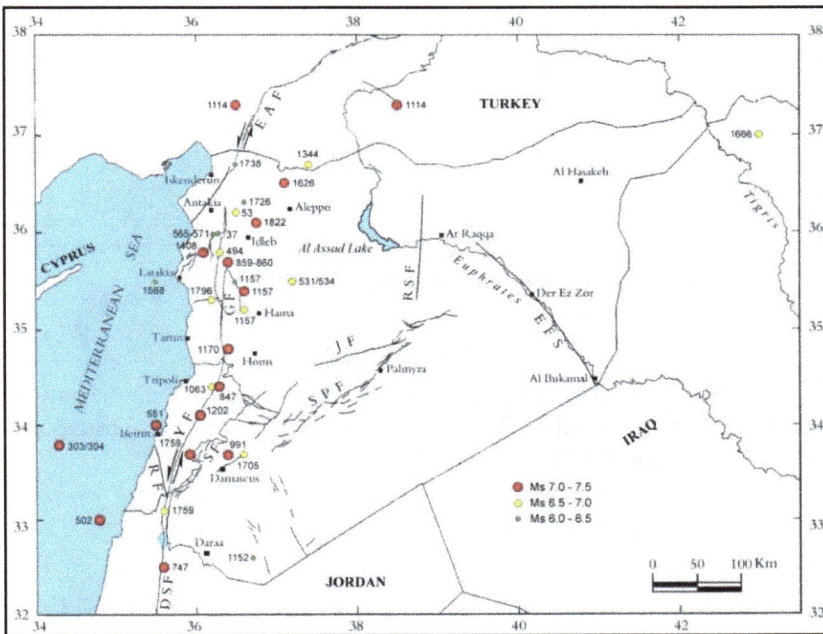

Figure 6.5: Distribution of Large Historical Earthquakes in and around Syria.

Instrumental Seismicity Studies

Using seismic data from international centres and regional bulletins, we compiled a catalogue of earthquakes occurred in Syria. Unlike the historical time, the seismicity of Syria is moderate during the last century with magnitude about 5°. The main instrumental seismicity is concentrated along the northern part of the DSFS and EAFS. Figure 6.6 shows the seismicity during 1900-1993.

Conclusions

The earthquake hazards assessment studies are ongoing in Syria to understand the earthquake activity and the behaviours of the faults. However, these multi researches should certainly contribute in mitigation the hazards in Syria and surrounding regions.

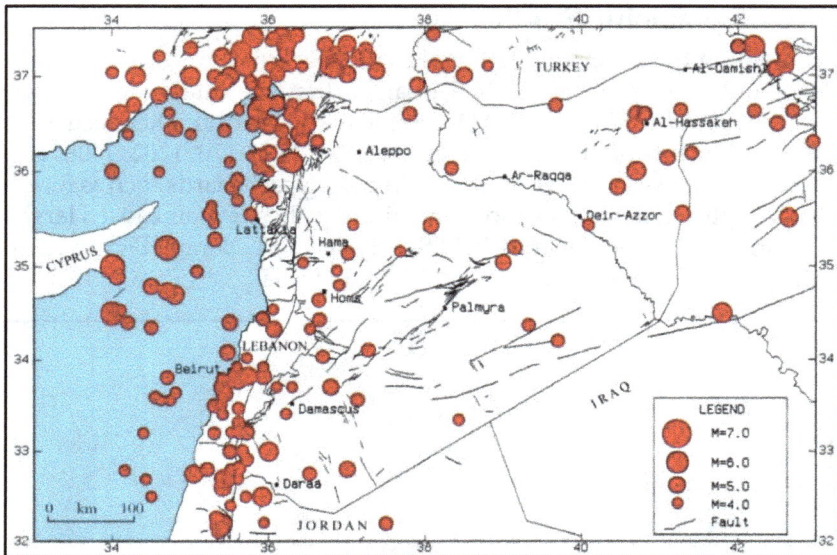

Figure 6.6: Distribution of the Seismicity of Syria during 1900–1993.

References

[1]   Dakkak, R., M. Daoud, M. Mreish and G. Hade [2005] "The Syrian National Seismological Network (SNSN): monitoring a major continental transform fault", *Seismological Research Letters* 76, 437-445.

[2]   Daoud, M. [2006] "Seismic Hazards evaluation for Syria", In: *The First Syrian Geological Symposium*, Damascus.

[3]   Meghraoui, M., F. Gomez, R. Sbeinati, J. Woerd, M. Mouty, A. N. Darkal, Y. Radwan, I. Layyous, H. Al Najjar, R. Darawcheh, F. Hijazi, R. Al-Ghazzi and M. Barazangi [2003] "Evidence for 830 years of seismic quiescence from palaeoseismicity, archaeoseismicity and historical seismicity along the Dead Sea fault in Syria", *Earth and Planetary Science Letters* 210, 35-52.

[4]   Gomez, F., M. Meghraoui, A. N. Darkal, F. Hijazi, M. Mouty, Y. Suleiman, R. Sbeinati, R. Darawcheh, , R. Al-Ghazzi and M. Barazangi [2003] "Holocene faulting and earthquake recurrence along the Serghaya branch of the Dead Sea fault system in Syria and Lebanon", *Geophysical Journal International* 153, 658-674.

[5]   Sbeinati, M.R., R. Darawcheh and M. Mouty [2005] "The historical earthquakes of Syria: an analysis of large and moderate earthquakes from 1365 B.C. to 1900 A.D.", *Annals of Geophysics* 48, 347-435.

[6]   Sbeinati, M.R., R. Darawcheh and M. Mouty [Field archaeological evidences of seismic effects in Syria", In: *Historical Investigation of European Earthquakes* 2, [P. Albini and A. Moroni, eds.], CNR, 195-203.

Chapter 7

Risks Posed by Large Seismic Events in the Gold Mining Districts of South Africa

R.J. Durrheim[1]*, R.L. Anderson[2], A. Cichowicz[3],
R. Ebrahim-Trollope[4], G. Hubert[5], A. Kijko[6], A. McGarr[7],
W.D. Ortlepp[8] and N. Van Der Merwe[9]

[1]CSIR and University of the Witwatersrand, South Africa
[2]California Seismic Safety Commission, U.S.A.
[3]Council for Geoscience, South Africa
[4]GeoHydroSeis c.c., South Africa
[5]Golder Associates (Pty) Ltd., South Africa
[6]University of Pretoria, South Africa
[7]U.S. Geological Survey, U.S.A.
[8]SRK Consulting, South Africa (deceased)
[9]University of the Witwatersrand, South Africa
*E-mail: rdurrhei@csir.co.za

ABSTRACT

Examining the occurrence of seismic activity in South Africa, the authors point out that the seismic event on 9 March, 2005 could be ascribed to past mining, and that seismic events will continue to occur in the gold mining districts as long as deep-level mining takes place and are likely to persist for some time even after mine closure. Seismic monitoring should continue after mine closure, and the seismic hazards should be taken into account when the future use of mining land is considered.

The national and local monitoring networks, operated by the Council for Geoscience and mining companies, respectively, are on a par with those installed in seismically active mining districts elsewhere in the world. However, steps should be taken to improve the quality of seismic monitoring and to ensure continuity, especially as mines change hands. The Klerksdorp and Free State gold mining districts are incorporating the risks of seismicity in their disaster management plans, and Johannesburg is urged to do likewise. Some buildings are considered vulnerable to damage by large seismic events, posing safety and financial risks.

It is recommended that an earthquake engineer inspect the building stock and review the content and enforcement of building codes. Appropriate training should be provided to all members of emergency services, and drills should be practiced regularly at public buildings to avoid panic should a large seismic event occur.

Introduction

A seismic event with a local magnitude M_L=5.3 occurred at 12h15 on 9 March, 2005 at DRDGold's North West Operations, Northwest Province, South Africa. The event and its aftershocks shook the nearby town of Stilfontein, causing serious damage to some buildings (Figure 7.1). Shattered glass and dislodged masonry caused minor injuries to 58 people. Three schools, two commercial properties, three blocks of flats, the civic centre and 25 houses suffered major damage. The direct cost of the damage to buildings in Stilfontein is estimated at R20 to R30 million.

Figure 7.1: Damage to Buildings in Stilfontein Caused by the M_L=5.3 Tremor on 9 March, 2005.

At the mine, two mineworkers lost their lives, and 3200 mineworkers were evacuated from the workings under difficult circumstances. The No. 5 Shaft and its infrastructure suffered severe damage, effectively halting mining operations in that section of the mining complex. The mine went into liquidation shortly afterwards, and some 6500 mineworkers lost their jobs. Approximately R500 million was claimed from insurers for damage to mine infrastructure and loss of production.

The events on 9 March, 2005 raised some wider questions. For example:

☆ Is the technology available to cope adequately with large seismic events in the current situation of remnant mining, deeper mines, and mining within large mined-out areas?

☆ Are current approaches to mine planning, design, monitoring, and management appropriate and adequate?

☆ Does mining, past and present, trigger or induce large seismic events? Will it continue to do so in the future?

☆ Can the effect of seismicity on mining towns and communities be limited and, if so, how?

Furthermore, other developments pertinent to the DRD situation and elsewhere raised an additional concern; namely, the relationship between mine water and seismicity.

In response to these concerns and in consultation with her principals at the Department of Minerals and Energy, the Chief Inspector of Mines, Ms May Hermanus, commissioned an investigation to assess the risks to miners, mines and the public arising from seismicity in mining areas, with particular reference to gold mines, remnant mining, pillar mining and mining districts in which mines are largely mined out or where flooding is occurring. The investigation team was instructed to draw on current knowledge and expertise to arrive at conclusions, make recommendations, and identify further research needs. While there was considerable discussion and debate between team members, it did not attempt to achieve complete unanimity on all points. The main report was endorsed by all team members, and represents a consensus view.

Mining and Seismicity in South Africa

Gold was discovered near present-day Johannesburg in 1886, and earth tremors and rockbursts were already a cause of concern to mining communities as early as the first decade of the 20th century. In 1908, minor damage in a village near Johannesburg led to the appointment of a committee, chaired by the Government Mining Engineer, to "inquire into and report on the origin and effect of the earth tremors experienced in the village of Ophirton". The 1908 Committee found that "under the great weight of the superincumbent mass of rock ... the pillars are severely strained; that ultimately they partly give way suddenly, and that this relief of strain produces a vibration in the rock which is transmitted to the surface in the form of a more or less severe tremor or shock" (Anon, 1915). At that time, there were no seismograph recordings that could be used to determine the magnitude of these earliest seismic events with any kind of scientific correctness. Following a recommendation of the 1908 Committee, a seismograph was installed at the Union Observatory in Johannesburg in 1910. Another seismograph was installed in Ophirton, but was moved to Boksburg in 1913, where shocks were beginning to be felt. It was only 50 years later that reliable seismic monitoring networks were established.

As the gold mining industry expanded and the severity of the problem increased, further committees were appointed in 1915, 1924, and 1964. From early on there was recognition of the need to distinguish between natural earthquakes and mining-related seismic disturbances. The 1915 committee was asked to "investigate and report on: (a) the occurrence and origin of the earth tremors experienced at Johannesburg and elsewhere along the Witwatersrand; (b) the effect of the tremors upon underground workings and on buildings and other structures on the surface; (c) the means of preventing the tremors." It was concluded, "the shocks have their origin in mining operations", and "while it may be expected that severer shocks than any that have yet been felt will occur in Johannesburg, their violence will not be sufficiently great to justify the apprehension of any disastrous effects" (Anon, 1915).

The 1924 committee was appointed "to investigate and report upon the occurrence and control of rock bursts in mines and the safety measures to be adopted to prevent accidents and loss of life resulting therefrom" (Anon, 1924). The committee extended its enquiries as far as Canada, USA, and even India, where large "area rockbursts" were believed to be associated with major faulting in the Kolar goldfield in Mysore State. The 1924 committee made many recommendations about general mining policy, protection of travelling ways, and the mining of remnants.

The 1964 committee was mandated to "study the question of rockbursts and to revise the recommendations of the Witwatersrand Rock Burst Committee (1924)" (Anon, 1964). The time was considered opportune as "not only had mining depths in excess of 11,000 feet below surface been reached on the Witwatersrand, but the rockburst danger had also revealed itself in the newer mining areas of the Far West Rand, Klerksdorp and the Orange Free State". The committee's recommendations were based on a considerable body of research and practical observations. The necessity for carrying out further research was noted.

Superficially, the key issues addressed by the recent investigation (Durrheim *et al.*, 2006) did not differ greatly from the issues addressed by the 1908, 1915, 1924 and 1964 committees. However, much has changed in the past four decades. The South African gold mining industry is mature and declining: production has been falling at an average rate of four per cent per annum for the past three decades. Only 342 tons of gold were mined in 2004, compared to 1000 tons in 1970. New problems are faced as mines approach the ends of their lives, cease operation, and workings are flooded. Meanwhile, many of the cities and towns in the gold mining districts have grown, and several seismic events with magnitudes exceeding 5 have caused damage to residential, commercial, and civic buildings in these towns. In addition, a great deal of rock mechanics and seismological research has been conducted since 1964. Much of it is reported in the proceedings of the six quadrennial *Rockbursts and Seismicity in Mines Symposia* held since 1982 and reviewed by Ortlepp (2005). Nevertheless, South Africa remained a major gold producer, the gold mines directly employed over 100,000 workers, and gold remained a significant source of foreign exchange. It is clearly in the national interest to ensure that the gold mines continue to generate these benefits as long as possible without posing unacceptable risks to mineworkers and the public.

Large seismic events causing serious damage to surface and/or mine infrastructure first occurred in the Free State region in the 1970s. An M_L=5.2 seismic event caused a six-storey apartment block to collapse in Welkom in December 1976 (Figure 7.2).

Figure 7.2: Damage to a Building in Welkom Caused by a M_L=5.2 Tremors on 8 December, 1976.

Fortunately, it was possible to evacuate the building before it collapsed. This event first evoked the idea of a natural earthquake being the ultimate cause. For example, Dr. Piet Pienaar, consulting geologist for Anglo American Corporation, was reported in the press to have said that the "Welkom earth tremor was almost certainly the result of geological phenomena and not caused by mining activities" (*The Friend*, 10 December 1976), and a report by foreign geotechnical consultants allowed this explanation to gain acceptance (W.D. Ortlepp, pers. comm.).

The media again used the term "earthquake" in 1989 when the "second Welkom earthquake" caused minor but widespread damage on surface. However, observed displacements on the President Brand fault in a nearby mine demonstrated that the origin of the earthquake was indeed close to mine workings. Another notable event that caused significant damage to surface buildings and mine infrastructure in the Free State region occurred in 1999, and became known as the "Matjabeng earthquake". It was similarly associated with conspicuous movement and some damage on a large fault (the Dagbreek), which extended across kilometres of contiguous mining.

The investigation team compiled a catalogue of damaging seismic events that have occurred in South Africa since 1900 by searching the seismological bulletins of the Council for Geoscience, scientific publications, mining company records, and newspaper articles (Durrheim *et al.*, 2006). The catalogue is not claimed to be complete or exhaustive, but is believed to list most mining-related events that have caused significant damage to buildings on the surface, or caused serious damage to mine infrastructure such as shafts or winding machinery that could have endangered many lives. In the 15-year period 1991-2005, the Council for Geoscience recorded 113 seismic events with $M_L > 4$ in the gold mining districts of South Africa (Table 7.1). However, only about 30 of these events are listed in the catalogue, which indicates that many events with $M_L > 4$ did not cause significant damage to surface structures. However, the March 2005 event in Stilfontein, which precipitated the 2005/6 investigation, was preceded by at least 14 large events in the Klerksdorp mining district in the previous 32 years that caused significant damage to surface buildings, or serious damage to important mine infrastructure that might have resulted in catastrophic loss of life.

An important parameter in the estimation of seismic hazard is the maximum magnitude (M_{max}), yet it is one of the more contentious as it is most likely a size of earthquake that has not yet occurred in the region under study. M_{max} for natural and mining-related earthquakes in southern Africa has been estimated to be in the order of 7.5 and 5.5, respectively (Shapira *et al.*, 1989). A value of M_{max} not exceeding 5.5 for mining-related events is supported by the data in Table 7.1, if the approximation $M_{max} \gg M_L$ (largest recorded event) + [M_L(largest recorded event)-M_L(second largest recorded event)] (Dunn, 2005) is applied.

Table 7.1: Large Seismic Events in the Gold Mining Districts, 1/1/1991 to 31/12/2005

Number of Events	East Rand	Far West Rand	Klerksdo RP	Free State
$4 \leq M_L < 5$	2	27	68	16
$M_L \geq 5$	0	0	2	2
largest	4.1	4.7	5.3	5.1
The 2nd largest	4.0	4.6	5.0	5.1

Source: Council for Geoscience, 2006.

Investigation Process

The investigation was officially announced on 18 October 2005, following the release of the report on the inquiry into the seismic event of 9 March 2005 (Department of Minerals and Energy, 2005). Interested and affected parties were invited to bring their concerns and relevant information to the attention of the Investigating Team, and hearings were held in the gold mining districts during November and December, 2005. The interviews were recorded and transcribed, and interviewees were given the opportunity to correct and amend the transcripts to ensure that their views were accurately and comprehensively captured. (The transcripts were not included in the final report, but rather served as aide-memoirs for interviewees who made written submissions, and for members of the investigation team to use in formulating their findings.) Further discussions with interested and affected parties took place in January 2006, when the two Californian members of the investigation team visited South Africa. Submissions were received from representatives of the following parties.

☆ *Gold mining companies*: AngloGold Ashanti, Gold Fields, Harmony, Simmer and Jack, President Steyn, and South Deep made oral submissions. AngloGold Ashanti, Gold Fields, and Harmony supplemented their oral submissions with written submissions, and DRDGold made a written submission.

☆ *Labour organisations*: The National Union of Mineworkers, Solidarity, and United Association of South Africa made oral submissions.

☆ *Department of Minerals and Energy*: Principal and senior inspectors from the Free State, Northwest and Gauteng regions made oral submissions.

☆ *Tertiary educational institutions*: Prof. Huw Phillips, head of the Mining Engineering Department at the University of the Witwatersrand made an oral submission, while Prof. Nielen van der Merwe, head of the Mining Engineering Department at Pretoria University was a member of the investigation team.

☆ Mine seismology researchers and consultants: Both oral and written submissions were made.

☆ Disaster management authorities in local and district municipalities: Oral submissions were made by officials in the Free State, Klerksdorp and Gauteng regions.

☆ Risk management, business continuity and disaster recovery specialists: Representatives of financial institutions with infrastructure in the Johannesburg CBD made oral submissions.

☆ In addition, several independent consultants and equipment manufacturers made oral and/or written submissions.

Findings and Recommendations

The nine questions posed by the Chief Inspector of Mines in the investigation team's terms of reference form the basis of this section. Each question is stated, followed by the investigation team's findings and recommendations, as well as a short summary of supporting evidence and a discussion of relevant issues. The full report (Durrheim *et al.*, 2006) includes the terms of reference of the investigation, a catalogue of large seismic events in South Africa since 1900, details of the investigations team's activities, the reports by team members, and written submissions by interested and affected parties. A summary of the report was published by Durrheim *et al.* (2007).

Cause of the 9 March, 2005 Event

Question 1

Can the seismic events that occurred on 9 March, 2005 at DRDGold's North West Operations be ascribed to mining; *i.e.* did mining past and/or present trigger the seismic events?

Findings

The magnitude 5.3 event and its aftershocks can be ascribed to past mining. The event was caused by rejuvenated slippage on an existing major fault, with the extensive mining activities in the region contributing most of the strain energy. The chance of these events being solely due to natural forces is considered to be extremely small.

Recommendations

Some insurance policies only cover damage due to natural seismic events and exclude damage caused by mining-induced events. Consequently, disputes concerning the cause of seismic events arise between insurance companies and property owners and business owners following damaging events such as the one of 9 March, 2005. It is recommended that the terms and conditions of insurance policies are reviewed in the light of scientific evidence that most seismic events in gold mining districts, large and small, are mining-related.

Discussion

The finding that the seismic events on 9 March 2005 are mining-related is based on both statistical and mechanistic considerations. It is incontrovertible that deep mining causes the rocks surrounding excavations to deform, and that the strained rock mass sometimes fails suddenly, either along a pre-existing weakness or by initiating a new rupture, causing the surrounding rock to shake violently. Everyone who made submissions on this point agreed that most of the seismic events observed in the gold mining districts could be ascribed to mining. However, opinions differed regarding the contribution of natural forces, particularly when the larger magnitude events (greater than magnitude 4, say) were considered. These large events invariably take place along pre-existing geological weaknesses, and sometimes, may occur several hundreds of metres away from mining activities.

The southern African region is relatively stable from the point of view of natural crustal seismicity, as it is remote from the boundaries of tectonic plates. Nevertheless, seismic events violent enough to cause damage to surface structures occasionally occur. The magnitude 6.3 and 7.0 earthquakes that occurred in the Western Cape (September 1969) and in central Mozambique (February 2006), respectively, are examples of such events. However, statistical analysis demonstrates that the number of potentially damaging seismic events is far greater in gold mining regions than in surrounding regions. In his report to the investigation, Dr. Kijko showed that the number of events exceeding magnitude 3 in the Klerksdorp mining region exceeds the average for South Africa outside gold mining districts by a factor of 700 (Durrheim *et al.*, 2006). Dr. Kijko used a threshold of 3 for statistical purposes, as very few events with magnitudes exceeding 4 have been recorded outside the gold mining areas. There is no known difference in geology or natural stress field between the gold mining districts and adjacent regions that can account for this.

Studies have also shown that the extent of mining close to large faults in the Klerksdorp and Free State districts can account for the displacement and slip area required to produce events exceeding magnitude 5 (Brummer and Rorke, 1990). In these districts, unmined and highly stressed fault-loss areas may extend for distances as great as 10 kilometres.

Events with a similar magnitude have occurred before in the gold mining districts, and hence the event of 9 March 2005 cannot be considered "totally anomalous", as claimed by a witness at the Inquiry (Department of Minerals and Energy, 2005). For example, magnitude 5.2 events occurred in the Free State and Klerksdorp gold mining districts in 1976 and 1977, respectively. It is interesting to note that an analysis of seismicity in the Klerksdorp region indicated a recurrence time for magnitude 5 events of about 20 years (Gibowicz and Kijko, 1994: 334-335), though it must be noted that changes in mine production affect recurrence times, as the amount of seismic energy released is approximately proportional to the amount of rock that is mined (Gay *et al.*, 1995).

Mine seismologists distinguish between induced and triggered events: an event is "induced" if mining activity is considered to have provided most of the energy that is involved in the event, and "triggered" if the energy contributed by the mining activity is small but sufficient to create instability. Many studies have shown that sufficient energy is provided by mining operations to induce seismic events. In his submission, Dr. Spottiswoode showed that extensive mining induces stresses considered to be capable of triggering seismic events at considerable distances from excavations (Durrheim *et al.*, 2006). Furthermore, Dr. McGarr showed that the dewatering of the rock mass during mining operations will tend to stabilise faults that might have been close to failure (Durrheim *et al.*, 2006). The Earth's crust is normally saturated with water, and the fluid pressure opposes the gravitational and tectonic forces that clamp and stabilise faults. Hence, the dewatering of the rock mass, all other things being equal, will increase the stability of the faults. As parts of the crust may have a low porosity and/or permeability, the strengthening effect of dewatering is difficult to quantify for a particular feature without detailed studies of the properties of the fault zone.

Information submitted to the inquiry held to determine the cause of the seismic event of 9 March, 2005 (Department of Minerals and Energy, 2005) showed that the main seismic event was located within a few tens of metres of the No. 5 Shaft Fault and the reef horizon. The reef in this region had been extensively mined in the past. The shaft pillar that was being mined in the year prior to the seismic event was more than two kilometres away from the focus of the M_L=5.3 event. The stress perturbation due to the shaft pillar mining would have been relatively small at the focus of the seismic event. Thus the extensive mining carried out more than a decade prior to the event was probably the major influence causing the event.

Much of the debate concerning the contribution of natural forces and mining to large seismic events seems to be driven by concerns regarding liability and compensation, rather than by scientific inquiry.

Likelihood that Large Seismic Events will Recur

Question 2

What are the probabilities of repeat occurrences in the same or other mining districts, especially where the mines are mature and large areas are mined out?

Findings

Seismic events will continue to occur in the gold mining districts as long as mining continues. Seismicity will decrease as production diminishes, and will slowly reduce to the background levels when mining ceases. There are distinct differences between the mining districts with respect to the maximum magnitude of events. The Klerksdorp and Free State districts tend to experience larger events than the East, Central, West and Far West Land. These differences are ascribed to differences in geological structure and mining layouts. Events with magnitudes exceeding 4 may cause some damage to buildings on the surface, while events with magnitudes exceeding 5 may cause serious damage. At current production rates, an event that exceeds magnitude 5 is likely to occur in the Free State or Klerksdorp district every twenty years or thirty years, on average. It is considered unlikely that events with magnitudes exceeding 5.5 will occur in the gold mining districts (Shapira *et al.*, 1989)

Recommendations

Seismic monitoring should continue after mine closure to determine when seismicity has reached background levels. The seismic hazard should be taken into account when considering the future use of mining land and the building codes applicable to any structures. Research should be undertaken to identify and delineate the most hazardous geological structures.

Discussion

Many studies have shown that seismic energy release is roughly proportional to the volume of mining, though this relationship may differ between regions, mines, and reefs. No unusual change in this relationship has been noted as mines reach the end of their lives. The narrow subhorizontal stops created by gold mining tend to close due to creep effects in the rock mass, allowing the stress that has been induced by mining to dissipate slowly. It is, however, necessary to take special precautions if highly stressed remnants are to be mined, especially if they are intersected by faults and dykes. Planning to remove clamping or bracket pillars alongside major faults must be particularly conservative. The ones should be on the mine owner to demonstrate there are good reasons to believe that the risk is, in fact, low.

Seismic Risks Posed by the Flooding of Mines

Question 3

What are the effects of flooding and the corresponding rising water levels on the stability of faults and other geological features?

Findings

If a fault is permeable to fluids, flooding can increase the pore fluid pressure on its surfaces, diminishing the effective stress clamping the fault, and decreasing its stability. Consequently, seismic events are likely to be triggered as the water level rises in mines that have been closed and allowed to flood. Seismicity will decrease once the water table stabilises. It is considered unlikely that any event triggered by a rising water table will have a greater magnitude than the events that occurred during mining.

Recommendations

Further research should be conducted into the relationship between seismicity and flooding. The possibility that rising and/or fluctuating water levels may trigger seismic events must be taken into account when the future use of mine land is considered, as well as in building codes applicable to any structures.

Discussion

The pumping of water from depth is costly, and thus there is a strong incentive for mines to be allowed to flood once the ore reserve has been exhausted. There is a concern that the flooding may trigger seismicity. Observations of mine flooding in Canada, India, Japan, Poland, and the partial flooding of Stilfontein, South Deep, and ERPM mines in South Africa, show that this is indeed the case. These observations have been supplemented by a theoretical analysis carried out by Dr McGarr (Durrheim *et al.*, 2006). These studies indicate that the flooding-induced seismic events are unlikely to have larger magnitudes than events that occurred during mining, and are likely to cease once the water table stabilises. However, this is a topic that merits further research, and monitoring of seismicity and water levels is recommended.

The Impact of Scale of Magnitude

Question 4

What are the effects of seismicity on inter-mine water plugs and water barrier pillars?

Findings

It is possible that a seismic event could cause movement on a fault surface transecting a water plug and/or water barrier pillar, open up a fluid pathway, and allow flow of water into populated mine workings. While it is unlikely that such an occurrence would become an uncontrollable inrush, the consequences could be disastrous, and so the risk must be seriously addressed.

Recommendations

If the flooding of old workings adjacent to current operations is being contemplated, a thorough survey of the mining geometry and geology must be conducted to ensure that the integrity of the rock mass comprising the water barrier pillar is not compromised, a risk assessment must be performed, water plugs must be designed according to internationally acceptable standards, and the seismicity and water flow must be monitored. Depending on the outcome of the risk assessment, water doors may be constructed and extra pumping capacity provided. Evacuation plans should be formulated and emergency drills conducted regularly.

Discussion

As noted above, the rise in water level on one side of a water barrier pillar is likely to be accompanied by an episode of seismicity. South deep mine was faced with the possibility of a water-filled compartment with a 1500 m head being created updip of its mining operations. The risk assessment and action plan implemented by South deep should be used as a benchmark for any other mine facing a similar situation.

Risk to Neighbouring Mines Posed by Large Seismic Events

Question 5

What are the seismic damage risks to neighbouring mines in areas in which mines are mature?

Findings

The risk of a seismic event on a mine causing damage to underground workings on a neighbouring mine depends on the distance between the focus of the event and vulnerable areas on the adjacent mines. There is some risk to mine workings within a kilometre or so of the mine boundary if a fault or dyke with a history of seismic activity transects the mining properties. The risk is not considered high because major infrastructure such as shafts are usually located a kilometre or more from mine boundaries and there is generally good cooperation between neighbouring mines with respect to mine planning and blasting schedules.

Capability to Manage the Risks Posed by Large Seismic Events

Question 6

What is the adequacy of current technology, monitoring, and communication methods in managing current risks?

Findings

A range of technologies, ranging from mining layout and sequencing strategies to local support methods, is available to mitigate the risks of underground damage resulting from large seismic events. However, there are some important reservations:

☆ Stope and tunnel support systems cannot guarantee the safety of workers within a few tens of metres of the source region of a large seismic event. Thus, if an area has already been extensively mined near any geological features that could host a large seismic event, any further mining of remnants or pillars adjacent to the structure must be cautiously considered and carefully planned.

☆ There is no technology anywhere in the world that is able to predict the time of occurrence of seismic events. The Council for Geoscience operates the South African National Seismic Network (SANSN) that monitors seismicity on a regional scale, while mines operate local monitoring systems. Both systems are considered useful and comparable with those installed in seismically active mining districts elsewhere in the world. However, steps should be taken to improve the quality and ensure the continuity of seismic monitoring, especially as mines change hands or close.

☆ The decline in seismological expertise on mines, at universities, and in research organisations during the past decade is a cause for concern.

Recommendations

☆ Additional SANSN stations should be established in each mining district in order to provide adequate seismic coverage. SANSN locations can be improved by using mine data to improve velocity models.

☆ Mine seismic monitoring systems and practice should be improved. For example: sensors able to determine accurately the source parameters of large events should be added; measures should be taken to prevent the loss of seismograms following power outages that large events often cause; the standard of maintenance should be improved; and seismic hazard assessment procedures should consider spatial dimensions and time scales appropriate for the occurrence of large seismic events.

☆ It is crucial that knowledge and technology continue to be developed through research work, accompanied by effective implementation and enforcement.

Discussion

Stope and Tunnel Support

It was noted that most mines have failed to implement the best of available support technologies for both stopping and tunnel development. In an endeavour to save immediate working costs, some mines have reverted to previously discarded support practices even where the latest understanding may deem safety to be compromised. It should be emphasised, however, that such deviations from best practice would have little or no influence on overall mine stability and the incidence of very large seismic events. The consequences of these practices are greater local damage to mine workings should a seismic event occur.

South African National Seismological Network (SANSN)

SANSN stations are distributed equally over South Africa. Consequently, stations located several hundreds of kilometres away are used to determine focal locations. The regional coverage is important for monitoring natural seismicity and to enable comparison with the mining districts. However, only events with magnitudes exceeding 2 are reliably recorded, and it is often impossible to locate the foci of these events with accuracy better than 5 km. While the mining district in which the event occurred can usually be indicated, it is difficult to determine the mine and impossible to identify the part of the mine or geological structure.

Mine Seismic Networks

All rockburst-prone mines have seismic networks that are useful for locating events, initiating rescue efforts, evaluating mining methods, and assessing seismic hazard. However, several shortcomings were identified:

☆ Only high frequency geophones (>4.5 Hz) are used. Consequently, the magnitude of large events cannot be accurately determined, as these large events radiate much of their ground motion at lower frequencies, leading to underestimates of M_L.

☆ Sometimes critical recordings of large events are lost as a result of power failure. It was recommended that uninterrupted power supply units be installed to prevent data loss, and that mine networks are synchronised with Universal Time, which will make it far easier to use data from adjacent mines in order to locate an event.

☆ Some mines have been lax in maintaining their seismic network, with the result that the quality of seismograms is poor.

☆ Some mines have failed to recalculate the seismic parameters of past events following the introduction of new software, with the result that there are discontinuities in the histories. Consequently it is difficult to assess trends and hazard. In some instances, seismic archives have been lost. This is especially a problem when mines have changed hands or closed.

☆ The spatial dimensions and time scales used to assess seismic hazard are sometimes unsuitable to determine the likelihood of occurrence of large seismic events.

Expertise

In the past decade there has been a serious decrease in high-level seismological and rock engineering expertise in South Africa. Practitioners have retired, emigrated, or changed occupation. Funding for research, which has not been generous in the past, has further diminished in recent years. Fundamental research into the source mechanisms of large seismic events and the damage mechanism of rockbursts is urgently needed.

Research Needed to Improve Stability

Question 7

Will the placement of slimes and other mining discards underground alleviate the situation?

Findings

There is no merit in placing slimes or backfill in areas that have been mined out in the past for the purposes of reducing the risks of large seismic events, as much closure as already taken place. The placement of slimes or backfill should be motivated by the need to provide local stability for current mining operations or for the disposal of surface waste.

Recommendations

Stabilising pillars may also be used to control seismicity, and should be considered in any mining strategy. More research of a fundamental nature is required to improve the understanding of how bracket pillars affect the source mechanism of large seismic events.

Discussion

There are several reasons the placement of slimes or backfill in old mine workings is not regarded as a worthwhile endeavour:

☆ Some stress has already been regenerated through total closure;

☆ It would be extremely difficult to ensure significant filling; and

☆ It could make any future mining (*e.g.* stabilising pillars, low-grade ore) more difficult, thereby sterilising remaining ore resources.

The use of stabilising pillars to limit the amount of closure and to clamp geological structures should be carefully considered when mining occurs in the vicinity of potentially hazardous geological structures.

Vulnerability of Infrastructure and Readiness for Disaster

Question 8

What are the implications for disaster management in mining districts in which seismic risk exists?

Findings

Disaster management officials in the district municipalities that cover the Klerksdorp and Free State gold mining districts are incorporating the risks of seismicity in their disaster management plans. However, this does not appear to be the case for the Johannesburg metropolitan municipality.

Recommendations

Johannesburg officials were urged to include seismic hazards in their disaster management planning. Appropriate training should be provided to all members of emergency services, and drills should be practised regularly at all public buildings (schools, hospitals, offices, etc.) so that panic can be avoided in a seismic event occur. There is considerable scope to improve the seismic hazard assessment process and emergency response, and a comprehensive set of recommendations is provided in the report.

Assessment on Economic Scale of Losses

Question 9

What are the risks to existing surface structures and infrastructure near major geological structures in mining districts? What precautions or restrictions should be applied to these?

Findings

Some buildings in gold mining districts are considered to be vulnerable to damage, and even collapse during seismic events, posing safety and financial risks. There are simple, relatively inexpensive measures that could limit damage and the risk of injury. Major financial institutions in the Johannesburg Central Business District (CBD) are concerned that disruptions to services even for a few hours following a seismic event could affect business continuity and cause losses, and are considering moving critical operations out of the CBD.

Recommendations

An experienced earthquake engineer should be contracted to inspect the building stock and review the content and enforcement of building codes. Officials from Johannesburg Metropolitan Municipality should meet with members of the CBD Seismic Interest Group to discuss concerns and develop strategies.

Conclusions

Large seismic events with local magnitudes as up to 5.5 are likely to shake the gold mining districts of South Africa as long as deep level mining continues, and for several years after mining ceases while the rock mass gradually stabilises. Although it is unlikely that these events will cause large-scale destruction in nearby towns, the largest events are capable of causing damage to buildings and infrastructure, and injuries to people. It is prudent to enforce appropriate building codes should any new structures be built near deep mines. It is recommended that a survey of public buildings such as schools or hospitals be conducted to determine if any reinforcement or modification is required. Furthermore, emergency response and disaster management agencies should include mining-related earthquakes in their response plans.

Acknowledgements

The South African Department of Minerals and Energy is thanked for permission to publish this paper. Many stakeholders in the gold mining districts contributed to the investigation, including representatives of labour, government and industry. Special thanks are due to those who made oral and written submissions, practitioners who compiled hydrological and seismological data, and those who offered hospitality to the Californian members of the team during their visit to South Africa. Ms Sue Kimberley, Mr Andre Mouton and Ms. Annette Joubert provided secretarial, bookkeeping and information services, respectively, to the investigation team. They are thanked for their help.

References

[1]   Anon (1915) Report of the Witwatersrand Rock Burst Committee 1915, The Government Printing and Stationery Office, Union of South Africa, Pretoria, 28 pp.

[2]   Anon (1924) Report of the Witwatersrand Earth Tremors Committee 1924, The Government Printer, Union of South Africa, Cape Town, 64 pp.

[3]   Anon (1964) Recommendations of the Rock Burst Committee, 1964, Department of Mines, Republic of South Africa, Cape Town, 12 pp.

[4]   Anon (1988) Risk Analysis and Management of Projects, Institution of Civil Engineers, Thomas Telford.

[5]   Brummer, R.K. and Rorke, A.J. (1990) Case studies on large rockbursts in South African gold mines. Proceedings of the 2nd International Symposium on Rockbursts and Seismicity in Mines, C. Fairhurst, (ed), Balkema, Rotterdam, pp. 323-329.

[6]   Department of Minerals and Energy (2005) Report in terms of section 72(1)(b) of the Mine Health and Safety Act (Act No. 29 of 1996) of the Inquiry into the seismic event which occurred on 9 March 2005 and resulting in the death of Tsepang Benjamin Matsoele and Tieho Mabe at No. 5 Shaft Complex, DRD Northwest Operations, Magisterial District of Klerksdorp in the North West Province.

[7] Dunn, M. (2005) Seismicity in a scattered mining environment – a rock engineering interpretation. Proceedings of the 6th International Symposium on Rockbursts and Seismicity in Mines, Potvin, Y. and Hudyma, M. (eds), Balkema, Rotterdam, pp. 337-346.

[8] Durrheim, R.J., Anderson, R.L., Cichowicz, A., Ebrahim-Trollope, R., Hubert, G., Kijko, A., McGarr, A., Ortlepp, W.D. and Van der Merwe, N. (2006) Investigation into the risks to miners, mines, and the public associated with large seismic events in gold mining districts, Department of Minerals and Energy, Pretoria.

[9] Durrheim, RJ, RL Anderson, A Cichowicz, R Ebrahim-Trollope, G Hubert, A Kijko, A McGarr, WD Ortlepp and N van der Merwe, (2007) The Risks to Miners, Mines and the Public Posed by Large Seismic Events in the Gold Mining Districts of South Africa, Chapter 4 in Challenges in Deep and High Stress Mining, Y Potvin, J Hadjigeorgiou and D Stacey (editors), Australian Centre for Geomechanics, ISBN 978-0-9804185-1-4, pp. 33-40.

[10] Gay, N.C., Durrheim, R.J., Spottiswoode, S.M. and Van der Merwe, A.J. (1995) Effect of geology, in-situ stress and mining methods on seismicity in Southern African gold and platinum mines. Proceedings of the International Congress of Rock Mechanics, Tokyo, Japan. Fujii, T. (ed.). International Society for Rock Mechanics, Vol. 3, pp. 1321-1325.

[11] Gibowicz, S.J and Kijko, A. (1994) An Introduction to Mine Seismology, Academic Press, New York, 399 pp.

[12] Ortlepp, W.D. (2005) RaSiM comes of age – a review of the contribution to the understanding and control of mine rockbursts, Proceedings of the 6th International Symposium on Rockbursts and Seismicity in Mines, Potvin, Y. and Hudyma, M. (eds.), Australian Centre for Geomechanics, Perth, pp. 3-20.

[13] Shapira, A., Fernandez, L.M. and Du Plessis, A. (1989) Frequency-magnitude relationships of Southern African seismicity, Tectonophysics, 167: 261-271.

Chapter 8
Cyclone and Wind Disaster Mitigation

N. Lakshmanan
Structural Engineering Research Centre,
CSIR Campus, Taramani, Chennai, India
E-mail: nlaxman@sercm.org

Introduction

Natural events such as cyclones, earthquakes and floods cause enormous damage in built-up areas leading to loss of life and properties. Society takes a long time to recover from such awesome disasters. The statistics of physical exposure to earthquakes, cyclones, and floods, the three major disasters worldwide, as published in "A Global Report – Reducing Risk: A Challenge for Development" by United Nations Development Program-2004, is given in Figure 8.1. The average annual deaths in these disasters are also given in the same report and are reproduced in the same figure. It is very clear that viewed in terms of frequency of occurrence and the extent of damage, cyclones should rank as the prime natural disaster in peninsular India. Approximately four to five cyclones form every year in the Bay of Bengal, and about one or two in the Arabian Sea. One or two cyclones cross the coastal regions of the country. During the decade of the eighties, the average annual loss due to cyclones was estimated as Rs.200 crores. Some of the major cyclones in the decade of the nineties, the International Disaster for Natural Disaster Reduction (IDNDR), namely the Kakinada Cyclone of Andhra Pradesh (1996), the Porbandar Cyclone of Gujarat (1998), and the Paradip Cyclone of Orissa (1999), caused economic damage varying between Rs.1000 crores to Rs.4000 crores. Table 8.1 gives the statistics of damage of a few major cyclones that crossed the Indian coasts.

Because of the improved facilities created by the IMD for tracking of the cyclones and timely warnings together with other infra structural facilities available such as cyclone shelters, there has been a drastic reduction in loss of life, but the damage scenario to buildings, structures and other infra structural facilities still presents a depressing scene. Though retrofitting of vulnerable structures has been suggested in

Figure 8.1: Physical Exposure and Relative Vulnerability to Various Natural Hazards.

Earthquakes

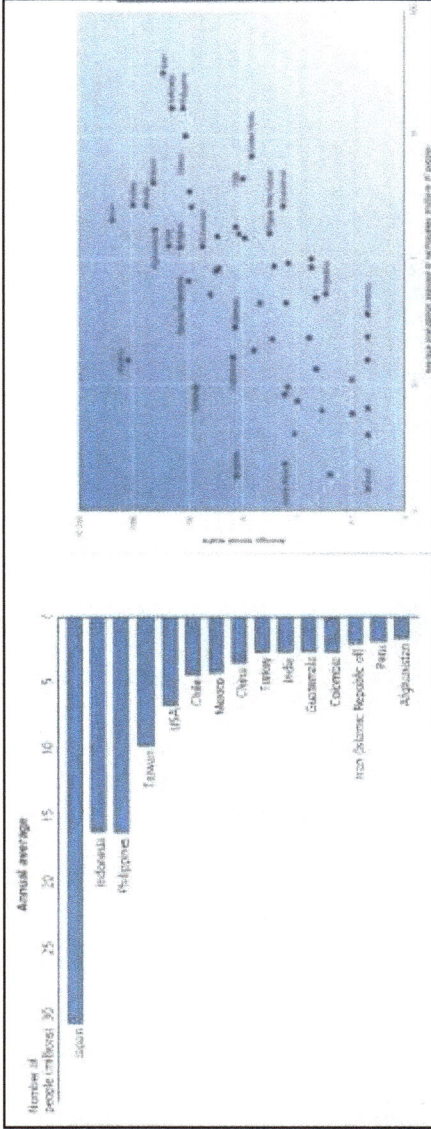

Cyclone

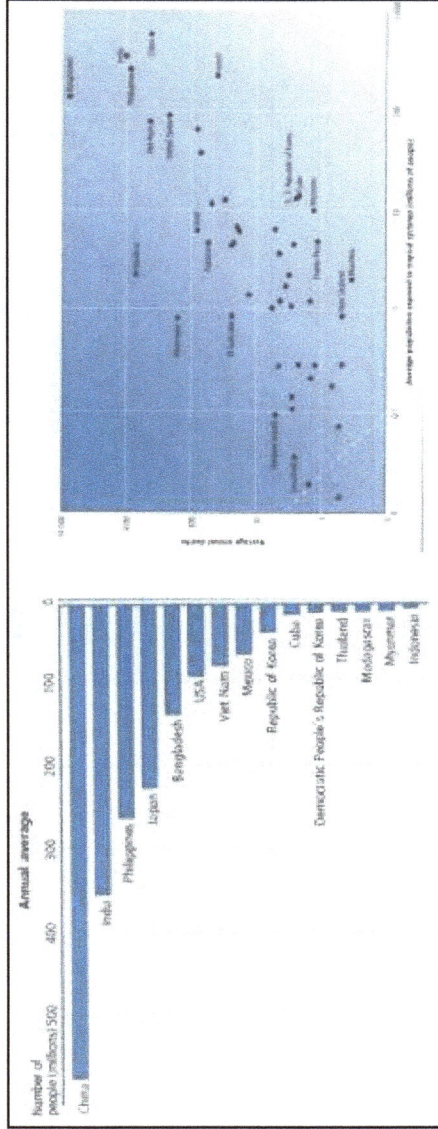

Contd...

Figure 8.1–Contd...

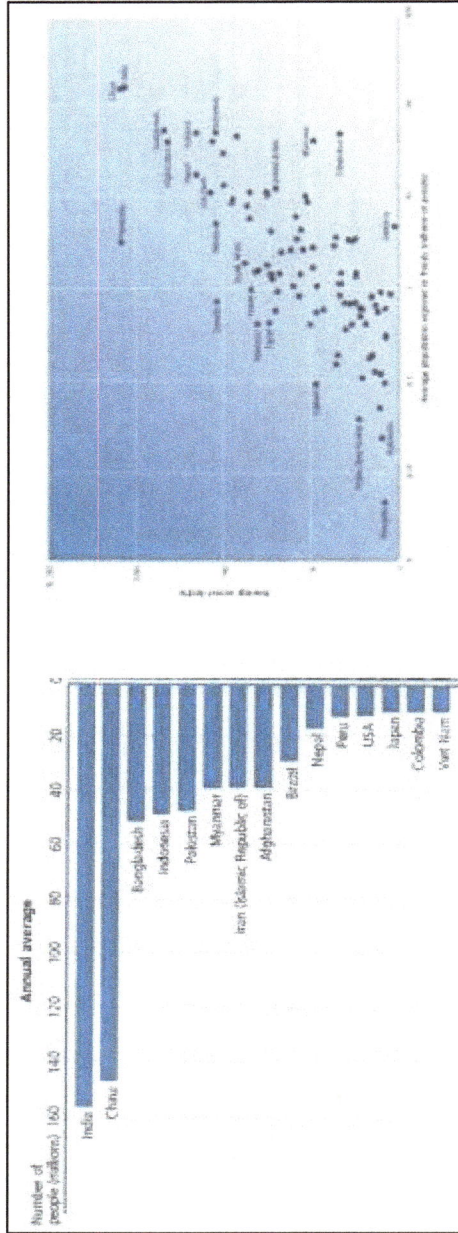

Floods

Table 8.1: Statistics of Damage due to Different Cyclones.

Sl.No.	Month and Year of Occurrence	States Affected	Population Affected	Humanities Lost	Houses Damaged/ Collapsed	Economic Loss (Rs. in crores)
1	Nov. 1977	Tamilnadu, Andhra	71,00,000	10,000	1,20,000 / 36,000	450
2	Nov. 1982	Orissa	–	–	–	110
3	Nov. 1988	West Bengal	19,00,000	500	1,50,000 1,20,000	N.A.
4	Nov. 1989	Andhra	5,00,000	48	1,00,000 20,000	–
5	Nov. 1993	Tamilnadu, Pondicherry	2,00,000	65	30,000	650
6	Nov. 1996	Andhra	70,00,000	749	3,50,000	5,375
7	June 1998	Gujarat	20,00,000	1,167	3,30,000	1,200
8	Oct. 1999	Orissa	1,26,00,000	9,615	16,50,000	1,462

many national forums, and guidelines have been prepared for new construction as also for retrofitting, these have not been implemented.

Tropical Cyclones and their Impact

Cyclonic Wind Field

Modelling of cyclonic wind field is an important part in simulation studies (Chan, 1971; Georgiou, 1983; Vickery and Twisdale, 1995a, Vickery and Twisdale, 1995b; Vickery, et. al, 2000; and Wills et. al, 2000). Sea surface roughness, land-sea temperature differential, the vertically oriented vorticity created due to coriolis forces, asymmetries in the fast moving cyclone, non-linear equations of motion, are a few of the parameters that lead to the complexities in the definition of cyclonic wind field. The section through the cyclone is shown in Figure 8.2. All numerical models emanate from the force balance equation given by

$$\frac{\delta P_r}{\delta r} = \rho\left(\frac{V^2}{r} + 2r\Omega\sin\Psi\right) \qquad \ldots (1)$$

where,

P_r: Tadial pressure

r: Pressure from the centre of cyclone

ρ: Density of moist air

V: Circumferential wind velocity at distance

Ω: Rotational speed of earth

and, Ψ: Latitude of place

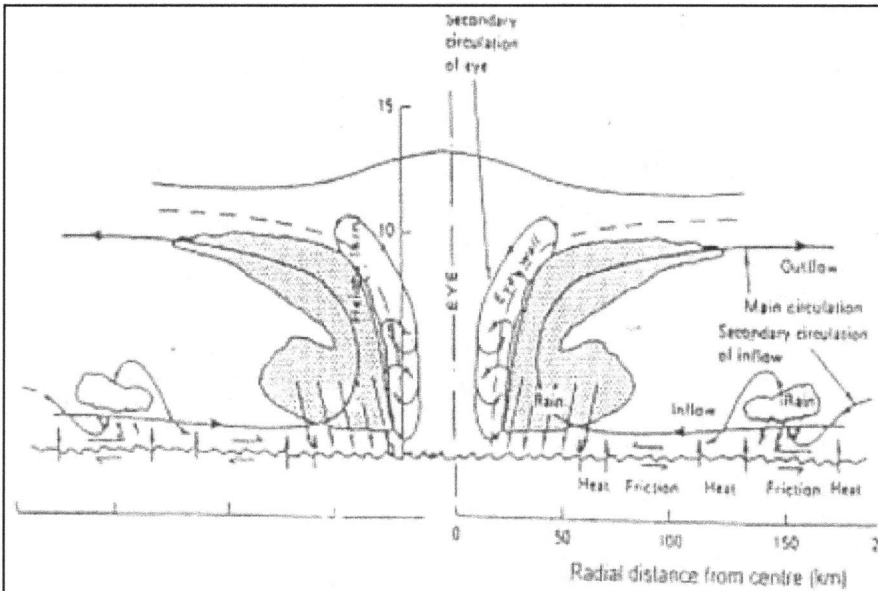

Figure 8.2: Section through Cyclone Wind Circulation.

The above equation is applicable only above the gradient height. One of the simplest forms of variation of V with radial distance has been given by Gomes and Vickery (1976)

$$(P_r - P_0) = (P_n - P_0)e^{-\left(\frac{Rm}{r}\right)}$$ ...(2)

where

P_0: Pressure at centroid of cyclone

P_n: Asymptotic pressure at large radius,

and, R_m: Radius at which maximum wind speed occurs

The normal procedure is to convert the wind speed at gradient height to that at 10 m level. Later log law model is used to obtain wind speeds in boundary layer region. For obtaining asymmetry, the local undisturbed wind speed is added on one flank and decreased on the other. Also transitional velocities are suitably added to obtain the full definition of cyclonic wind field. It is important to recognize that for damage estimation what is required is not the temperoral definition of wind field, but an integrated, whole time, spatial wind field effects at different locations. To realize such objectives reasonable approximations can be made. Variation in cyclonic wind speed with radial distance can be expressed as (Figure 8.3).

$$V_r = V_0\left(\frac{R_m}{r}\right)^\alpha$$ ...(3)

where

V_r: Wind speed at radial distance, $r \geq R_m$

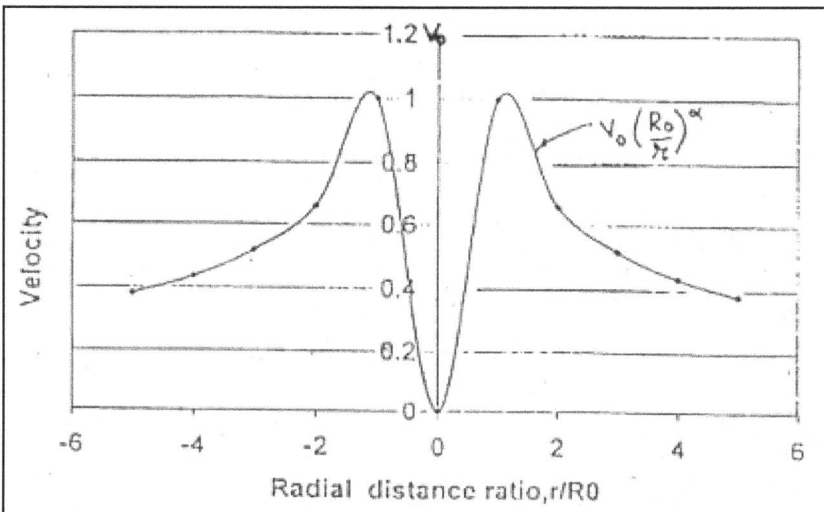

Figure 8.3: Variation of Wind Speed with Radial Distance.

V_0: Maximum wind speed at radial distance R_m

and, α: a power law coefficient varying in the range of 0.5 to 0.67.

Considering the idealized transit of the cyclone over the shore line, and hypothesizing that the intensity of the cyclone does not reduce for a distance of 30 km from the shore line, idealized spatial wind field for damage estimation can be derived as shown in Figure 8.4 (Lakshmanan and Shanmugasundaram, 2002).

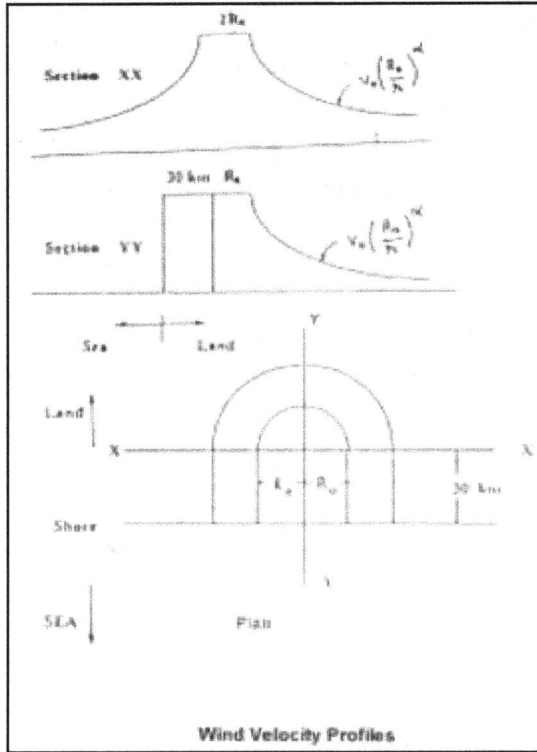

Figure 8.4: Cyclonic Wind Field for Damage Assessment.

Indian Initiatives in Cyclone Disaster Mitigation (Prem Krishna, 2004 and Lakshmanan 2006)

Organisational Efforts

A look at the Indian scene on initiatives for disaster mitigation would clearly reveal that the attention given to cyclone disaster mitigation is far less compared to earthquakes. Nevertheless, there have been some significant developments over the years. A Wind Engineering Society has been established in 1993 and the 9th International Conference on Wind Engineering was organised in 1995. Since then, bi-annual national conferences have been regularly organised by the Wind Engineering Society. Excellent facilities have been established for wind engineering research at the Structural Engineering Research Centre, Chennai, Indian Institute of Technologies (IITs), Roorkee; Kanpur; and Delhi. These centres have been carrying out R&D activities aimed at safe, reliable, and economical structures against wind loading. Visweswaraya National Institute of Technology, Nagpur, and IIT, Madras, have carried out some analytical investigations in the area of wind engineering. Indian Meteorological Department monitors weather system and is primarily responsible for cyclone warning bulletins. The IMD network has sophisticated equipment and facilities for this purpose.

Codes and Standards

India is a partner in regional harmonisation of wind loading and wind environmental specifications in Asia-Pacific Economies (APEC-WW 2004, 2006). The following are the major Indian standards to wind loading:

☆ IS:875-1987, Part:3, Indian Standard Code of Practice for Design Loads for Buildings and Structures, Wind Loads, BIS.

☆ IS:15498-2004, Guidelines for Improving Cyclonic Resistance of Low-Rise Houses and other Buildings and Structures, BIS.

☆ IS;15499-2004, Guidelines for Survey of Housing and Buildings in Cyclone-Prone Areas, BIS.

A draft revision for IS: 875-1987, Part 3 is under discussion. This would make the above standard in line with contemporary Codes of Practice existing worldwide.

Full-Scale Field Measurement of Wind Speeds and Dynamic Response of Structures

The cyclonic speeds were computed using the failures of some simple structures. Also detailed failure analysis were carried out on water towers, lattice towers, industrial sheds, lighting poles, hoardings and a number of other structures. It became evident then that extreme wind speeds alone cannot be the reason for failures of some of the structures. Failures of some of the structures were suspected to be due to dynamic action (Venkateswarlu *et al.*, 1995). Hence, it was felt essential that well designed full-scale experiments are needed to establish the cyclonic wind characteristics.

Full-scale experiments on measurements of wind characteristics and response of various buildings and structures are important as they provide valuable information for designers and scientists in better understanding dynamic wind effects on structures. However, at present in our county only very few such studies have been carried out.

Many engineers believe that the wind speed given by IMD is adequate for design of structures. They fail to realize that all weather system predictions are at least 1000 m above ground level, and as of now no civil engineering structure is that tall. Repeatedly it has been mentioned in many forums that well designed full-scale experiments are essential to establish cyclonic wind characteristic near ground level. Many researchers worldwide have recognized that the velocity profile of cyclonic winds and their turbine characteristics would be very different from normal pressure system winds. There have been attempts made at different times for a nationally coordinated effort in this direction. However, these have not borne the desired fruits. The wind energy sector has realized the need to conduct wind speed measurements, essentially to know the potential of wind energy and to develop the wind rose diagrams. Even they have failed to realize the need to capture cyclonic wind characteristics. A large number of wind energy support towers failed during the Gujarat cyclone of 1998. Based on long term and short term circulation models, wind power density for forecasting wind energy resources has been suggested (Deb et al, 2004). Wind-induced fatigue damage is extremely important for dynamically sensitive structures and a few cycles of large amplitude stress reversals are adequate to damage a structure. A well coordinated test programme is essential and should be taken up immediately at the national level.

Cyclonic Wind Speed Map

The intensity of damage to buildings and other infrastructural facilities led to a closer examination of the wind load provisions given in IS:875 (Part-3). While preparing the basic wind speed map of India, the Bureau of Indian Standards has not distinguished between the peak velocity due to well formed weather systems and cyclones. In order to estimate the cyclonic wind speeds rationally, cyclonic wind speed data pertaining to the coastal regions of Tamil Nadu, Andhra Pradesh, Orissa, West Bengal, Konkan and Saurashtra since 1891 were collected from IMD, and these data were augmented using Monte Carlo Technique. Only wind speeds exceeding 115 kmph that are known to cause damage to structures were considered. Based on Gumbel Type-I distribution the cyclonic wind speed having a return period of 50 years was computed. The cyclonic wind speed map so obtained is shown in Figure 8.5.

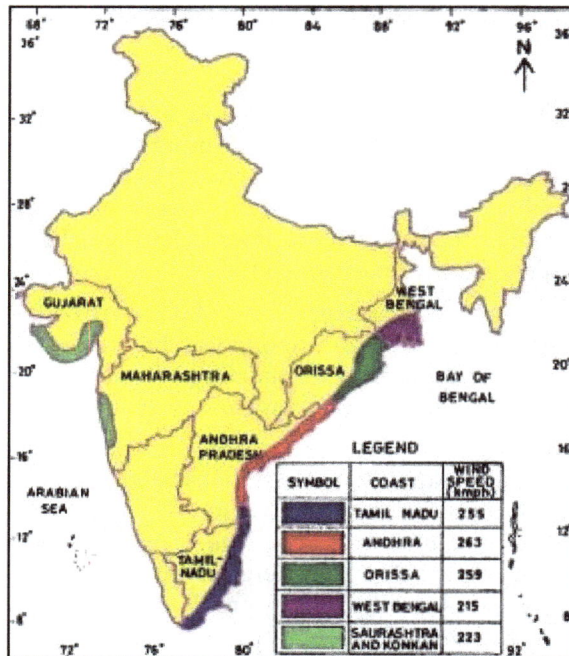

Figure 8.5: Cyclonic Wind Speed Map of Coastal India.

Considering cyclones as accidental occurrence and hence can be considered as equal to ultimate wind speed, the wind speed corresponding to working/service load stage was obtained. These were discussed in a number of workshops and seminars. A basic cyclonic wind speed of 65 m/sec. has been recommended for adoption in cyclone-prone regions. To obtain the risk coefficient k_1 the values of the parameters A and B defined in IS:875-1987 Part:3 were suggested as 95 and 35 kmph, respectively. Earlier SERC was also involved in developing the codal provisions in IS:875-1987, Part:3, particularly with respect to the evaluation of the risk factor k_1 and the terrain roughness and structure size factor k_2.

The Guidelines for improving cyclonic resistance of low-rise houses and other buildings and structures (IS: 15498-2004) slightly modified the above and the design wind speed in cyclone-prone regions is defined as

$$v_d = f.k_1.k_2.k_3.v_b$$

where, f is the enhancement factor equal to unity for houses, 1.15 for industrial buildings and 1.3 for buildings of post-cyclone importance. v_b in this case corresponds to the basic wind speed given in IS:875-1987 (Part 3).

Extreme wind quantile estimation using limited duration wind data over number of stations, their inter-distances, and using pooled frequency analysis approach has been attempted by Goel *et al.* (2004). The extreme wind estimates show significant deviation from wind speed map of the country included in IS:875-1987 (Part 3). There is large inconsistency such as Chennai having an extreme wind speed of 96 kmph as against Coimbatore showing 142 kmph. Similarly, in between Kakinada and Narasapur the wind speed is shown as 99 kmph. The wind speed evaluated in Saurashtra and Konkan Coasts are higher than that in the east-coast in many places. Since the ratio between peak gust speed and mean wind speed vary from station to station, further studies are necessary to tune this approach for wider acceptance.

Cyclonic Damage Surveys

The type of inadequacies in structural schemes, detailing practices, construction methodologies etc., led to large damage to life lines, residential buildings and industrial structures during the 1977 cyclone of Andhra Pradesh. In one of the subsequent cyclones, severe distress was noticed to water towers, and one of them even collapsed and this led to the need for looking more closely to the structural engineering aspects relating to cyclone induced damage. SERC realized the importance of "learning from failures", and all cyclones that caused major damage were followed by a post-cyclone damage survey by SERC. Table 8.2 gives a list of cyclone damage surveys and the damage observed to various buildings and structures.

Turbulent Characteristic of Cyclonic Winds

SERC has developed expertise for conducting reliable full-scale field experiments, which include instrumentation, measurement and analysis of test data. The experiments conducted on 101 m tall microwave tower are comparable to similar full-scale field experiments conducted anywhere in the world. Sophisticated instrumentation facilities were planned, procured and successfully installed on a number of full-scale structures, which included a low-rise roof, and a number of lattice towers of different heights.

Table 8.2: Cyclone Damage Surveys Conducted by SERC, Madras.

Year of Occurrence	Location of Land Fall	Damage Observed
1977	Machilipatnam, A.P., India	Life Lines, Residential Buildings, Industrial Structures, etc.
1977	Nagapatnam, T.N., India	Life Lines, Residential Buildings, Industrial Structures, etc.
1984	Sriharikota (SHAR), A.P., India	Roofing, Elevated Water Tanks, Communication Towers, Industrial Structures
1989	Kavali, A.P., India	Microwave Towers, Large Industrial Structures and Dwellings.
1990	Guntur, A.P., India	Residential Buildings, Life Lines.
1992	Miami, U.S.A.	Residential Buildings and Communication Lines.
1993	Near Karaikal, T.N., India	Industrial Structures, Residential Buildings
1994	Madras, T.N., India	Lamp Masts, Hoardings, Dish Antennae.
1996	Kakinada, A.P., India	Residential Buildings, Transmission and Communication Towers, Lamp Masts, Industrial Structures, etc.
1998	Near Porbandar, Gujarat, India	Residential Buildings, Communication Towers, Lamp Masts, Industrial Structures, Port and Marine Structures.
1999	Near Pradip, Orissa	Residential Buildings, Transmission and Communication Towers, Industrial Structures, Port and Marine Structures.

Extensive data was collected from a 52 m tall lattice tower, erected near Wind Engineering Laboratory at SERC, which had continuous wind monitoring under normal and cyclonic wind conditions. These have been analysed using special purpose software developed in-house. For the first time, the wind engineering group has been able to capture the spectra of cyclonic winds which has shown energy even up to 10 Hz. The cyclonic wind spectrum experimentally obtained is shown in Figure 8.6. This may explain the collapse of some of the well-engineered structures during cyclones due to dynamic action. A number of papers have been published based on the full-scale experiments conducted and a few of these are included under references.

Prediction of the Track of a Cyclone

It is very difficult to predict the track of any cyclone based on theoretical model alone because of the influence of various real time changes in meteorological and geological parameters. However, past data on the tracks of cyclones can be used to carry out probabilistic analysis for predicting the tracks of cyclones. For this purpose, data on tracks of cyclones which formed in the Bay of Bengal and Arabian Sea and crossed the Indian coast since 1981 have been obtained from IMD in a graphical form. These tracks have been digitized and stored in the Computer as a database. A software has been developed to obtain:

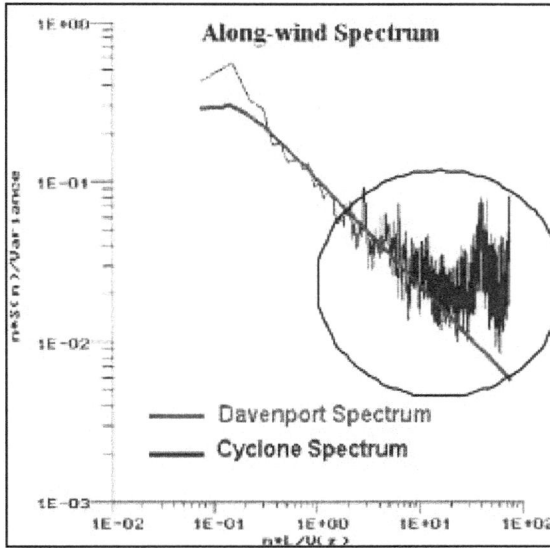

Figure 8.6: Cyclonic Wind Spectrum.

1. The probability of a cyclone crossing different cities along the coast of India for any particular month, from any point of origin.
2. To obtain the most probable path of a cyclone in a graphical form for a given place of origin and month.

Prediction of the cyclonic track and probable location of crossing, by using the software developed will help the community at large, in preparing themselves for facing the cyclone. The industries in the cyclone prone regions can benefit by obtaining the statistical probability of cyclone striking at that particular place, to plan their construction activities. IMD authorities have also appreciated the capabilities of the software developed given the location of formation of a depression.

The above software could predict well the track of the October 1999 Orissa cyclone and its place of crossing three days before the cyclone struck the coast of Orissa (Harikrishna, *et al.*, 2002).

Damage Modelling

Damage estimation is essential for proper planning of disaster mitigation measures. A simple analytical model has been proposed. The damage caused due to Orissa cyclone is evaluated and compared with actual field data. Damage to buildings and structures can be considered proportional to the power of wind. All buildings depending on their size, shape and type of construction have a maximum wind velocity which they can withstand without failure. This may be called as capacity wind speed. However due to many factors such as inherent deficiencies in layout, ageing, poor maintenance, improper detailing, poor quality of materials used and similar other factors the capacity wind speeds gets reduced. The reduced capacity wind speeds alone are to be considered in damage analysis.

In a developing country like India, non-engineered and semi-engineered buildings coexist with engineered buildings. The types of buildings available are also numerous. The Vulnerability Atlas of India prepared by an expert committee set-up by Government of India and published by Building Materials and Technology Promotion Board has categorized the building stock into nine categories (1997). For the purpose of the present analysis these have been regrouped into three categories namely non-engineered, semi-engineered and engineered buildings. The damage model proposed is verified with the actual damage that occurred during Orissa Super Cyclone.

Consider a damage index D based on the power of wind as given by

$$D = \left(\frac{V}{V_0}\right)^3 \qquad \qquad \dots (4)$$

where,

    V: Wind velocity in a region, and

    V_0: Reduced capacity wind speed.

When the reduced capacity wind speed is less than V, it is apparent that the building would be destroyed. However, if V_{0c} is greater than V various levels of damage are possible. It is hypothesized that no damage to buildings would result of the wind speed V is less than $0.5\,V_{oc}$, and a linear variation between damage indexes is proposed with the non-dimensional parameter (V/V_{oc}). As can be seen from Figure 8.7, the linear approximation proposed gives good estimates of damage index when compared to the cubic variation. Hence,

$$\left(\frac{V}{Voc}\right) = \frac{1}{2}(1+D) \ \text{ for } \ 0.5 \le \frac{V}{Voc} \le 1.0$$

$$D = 0 \qquad \qquad \text{for } \left(\frac{V}{Voc}\right) \le 0.5 \qquad \dots (5)$$

The value of V can be determined from the idealized integrated spatial wind field defined earlier during the transit of a cyclone (Figure 8.7).

Capacity Wind Speed

The estimates of capacity wind speeds for different categories, particularly those belonging to non-engineered and semi-engineered category poses problems. No engineering properties of materials used would be readily available. Also the scatter could be very high. It is very difficult to estimate the flexibilities provided at connections, particularly those due to use of organic ropes. There are eccentricities in connections. Hence it was decided to conduct full-scale tests on a typical building configuration using different type of roofs (Lakshmanan, *et al.*, 2003). The full capacity wind speeds for some of the roofs tested are given below:

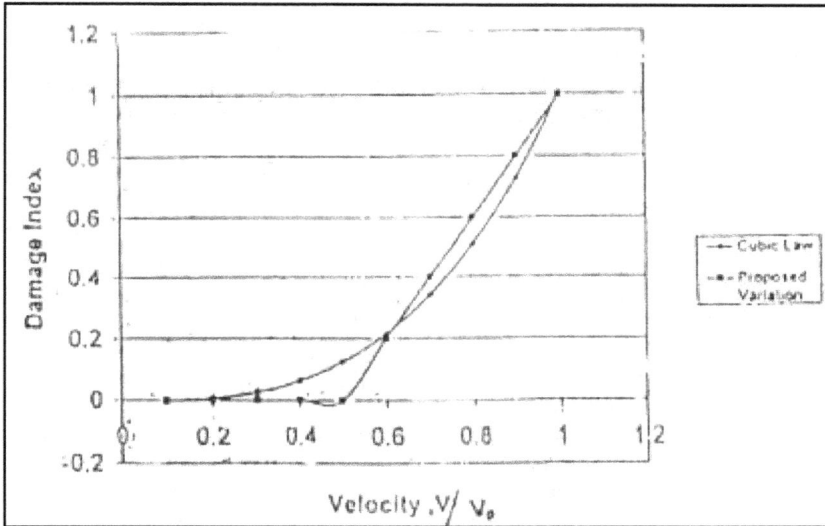

Figure 8.7: Variation of Damage Index with Non-Dimensionalised Velocity.

Table 8.3: Capacity Wind Speeds for Buildings.

Type of Building	V_{om}, m/sec
Conventional Gable roof	38
Gable roof with strengthening	50
Hipped roof with strengthening	58
AC sheet roofing with U-bolts	72

As mentioned earlier, the reduction in capacity wind speed can only be estimated by the data of structures on the ground. This can best be done by conducting a survey of buildings and structures in the cyclone prone regions. A detailed questionnaire for collecting data on buildings in villages located in the cyclone prone coastal regions of India was evolved (Lakshmanan and Shanmugasundaram, 1997). Data relating to the general parameters indicated in Table 8.4 were gathered and tabulated (Narayanan, *et al.*, 1997).

Table 8.4: Evaluation of Inherent Weakness in Building Category.

Data on	Parameters
Building details	Building type, age of building, quality of construction, maintenance, plan shape, height of plinth, percentage of opening.
Foundation	Type of sub-grade, type of foundation, material of foundation, depth of foundation, anchorage of foundation.
Walls	Material of wall, slenderness ratio, edge clearance to opening, anchorage of wall to foundation.
Roofs	Type of roof, roof slope, structural scheme, material of the roof, anchorage of the roof.

Based on the exhaustive analysis of the data, it was found that the fractional capacity reduction factors varied between 0.75 to 0.85 for non-engineered buildings, 0.80 to 0.87 for semi-engineered buildings for engineered buildings. It is common knowledge that long duration cyclones, inflict greater damage than short duration cyclones, and the reduced capacity wind speed was further factored by $[1/T^{0.02}]$ where, T is the duration of cyclones in hours. Using the above approach, the reduced capacity wind speeds in m/sec were worked out as given in Table 8.5.

Table 8.5: Reduced Capacity Wind Speed V_{oc}.

Type of Building	Cyclone 10 hrs.	Duration 24 hrs.
Non-engineered buildings	42	41
Semi-engineered Buildings	53	52
Engineered buildings	74.5	73

The above reduced capacity wind speeds together with the spatial wind field defined earlier with a value for α given in Figure 8.3 taken as 0.6, gives the radius of the damage field for different damage coefficients as given in Table 8.6. The peak wind speed during the Orissa Cyclone was taken as 75 m/sec. The damage zones are indicated in Figures 8.8–8.10.

Table 8.6: Definition of Damage Zones.

Sl.No.	Zone	Damage Index
1.	Zone 1	1.0
2.	Zone 2	0.8 – 1.0
3.	Zone 3	0.6 – 0.8
4.	Zone 4	0.4 – 0.6
5.	Zone 5	0.2 – 0.4
6.	Zone 6	0.0 – 0.2

Estimation of Damage

Using Figures 8.8–8.10, the area in each zone for different categories of buildings can be computed. The density of housing based on data available in vulnerability atlas and assuming a growth rate of 2 per cent per annum was computed as 43, 10 and 4.5 respectively for non-engineered, semi-engineered and engineered buildings respectively. It is common knowledge that the damage is not uniform in individual zones. For example, damage zone-1 does not suggest that all the buildings would be raced to the ground. This is because of a number of factors. One of them of course is that wind speed variation does have an asymmetry. Engineered buildings, for example, have an over strength factor, and alternate flow paths. SERC has conducted a number of cyclone damage surveys since the year 1977. Buildings getting affected on one side of the road, discrete buildings remaining totally unaffected, reinforced concrete roof slab being blown off which is highly unlikely, interior regions getting affected more than the coastal region, the non-uniform effects of flooding etc., have all been seen in the cyclone damage surveys. Hence use of a damage probability matrix as suggested for seismic damage evaluation is suggested. Damage probability matrices (DRM) can be developed based on the cyclone damage surveys. This is presently done based on expert judgment, and the damage probability matrices vary for different categories of buildings. The damage probability matrices help in correct estimates of damage to

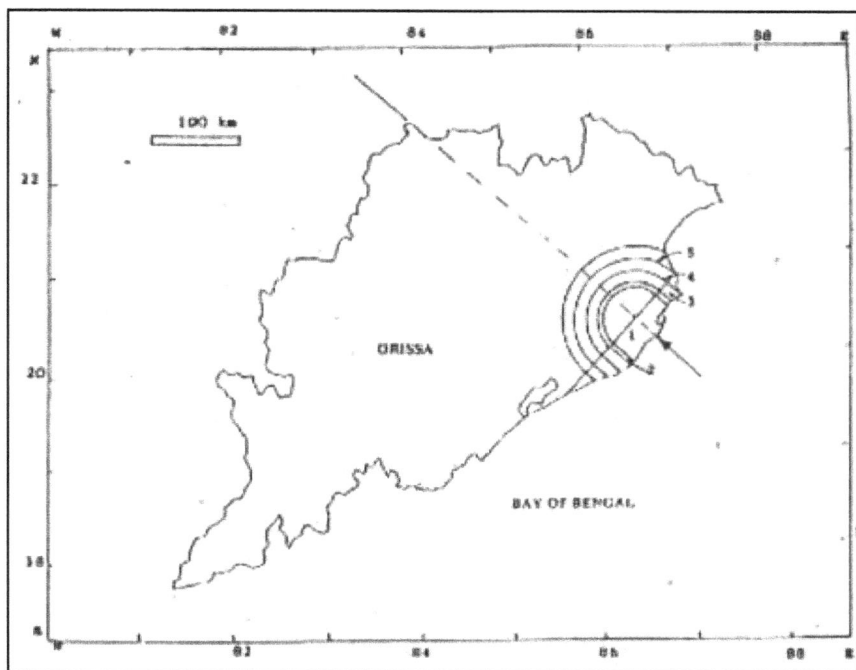

Figure 8.8: Damage Zone for Engineered Structures.

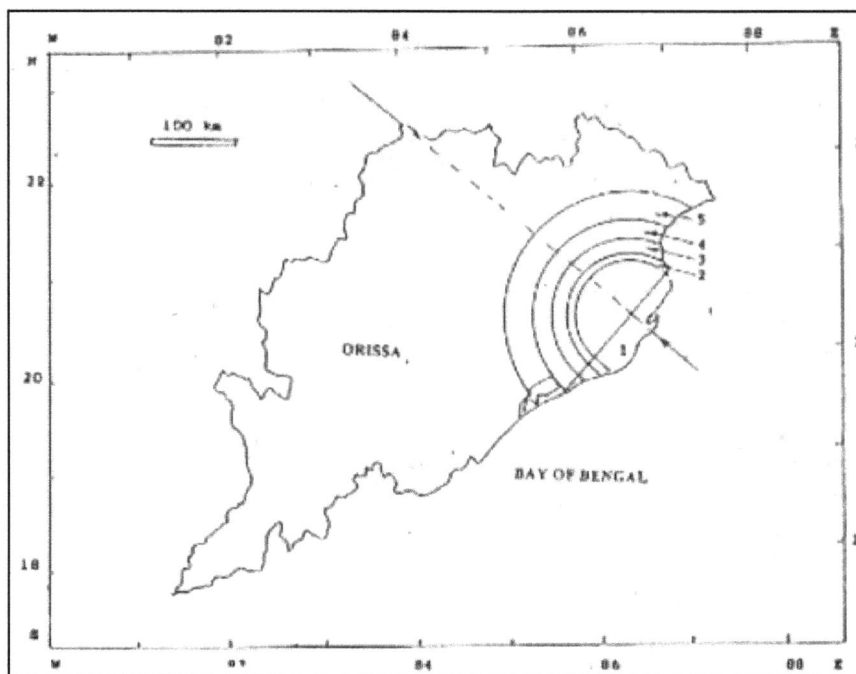

Figure 8.9: Damage Zone for Semi-Engineered Structures.

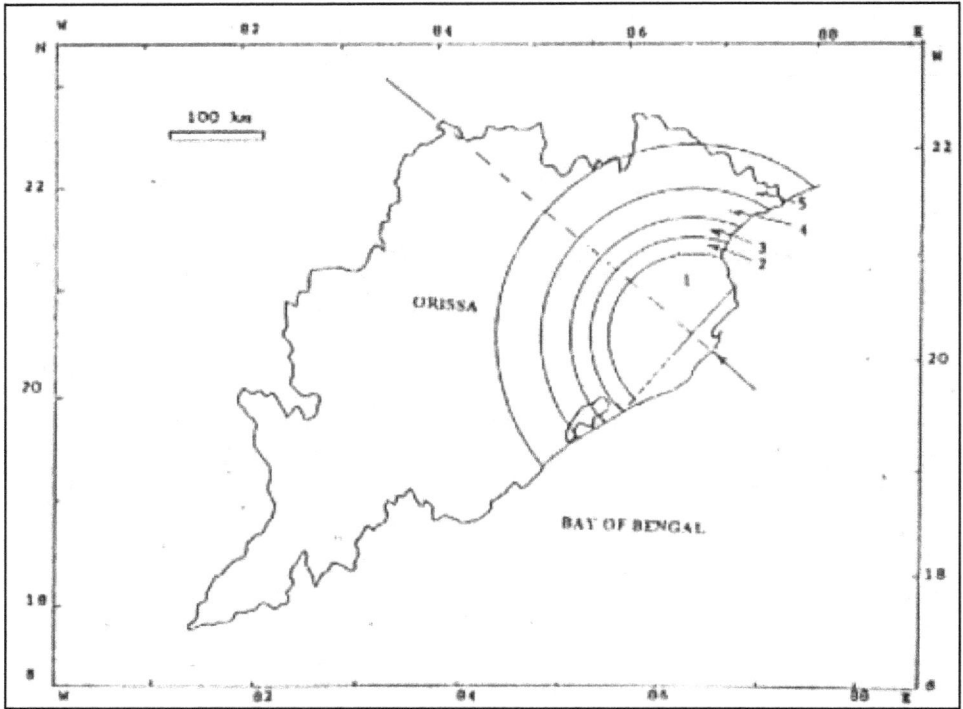

Figure 8.10: Damage Zone for Non-Engineered Structures.

buildings and structures. The damage probability matrices used for various zones, and for various categories of buildings are given in Tables 8.7–8.9. Low and very low levels of damage in zones 5 and 6 are not considered in further analysis as their economic and societal impact is marginal.

Table 8.7: DPM for Non-Engineered Buildings.

Type of Damage	Zone-1	2	3	4	5.	6
Destroyed	0.5	0.10	0	0	0	0
Very high	0.3	0.60	0.10	0	0	0
High	0.15	0.20	0.70	0.10	0	0
Medium	0.05	0.10	0.10	0.80	0	0
Low	0	0	0.10	0.10	0.9	0
Very low	0	0	0	0	0.1	1.0

Using the damage probability matrices and the number of non-engineered, semi-engineered and engineered buildings included in various zones, the damage has been estimated, and is given in Table 8.10.

Table 8.8: DPM for Semi-Engineered Buildings.

Type of Damage	Zone-1	2	3	4	5	6
Destroyed	0.50	0.05	0	0	0	0
Very high	0.25	0.60	0.05	0	0	0
High	0.15	0.20	0.70	0.05	0	0
Medium	0.10	0.10	0.15	0.80	0	0
Low	0	0.05	0.10	0.10	0.9	0
Very low	0	0	0	0.05	0.1	1

Table 8.9: DPM for Engineered Buildings.

Type of Damage	Zone-1	2	3	4	5	6
Destroyed	0.40	0.05	0	0	0	0
Very high	0.30	0.50	0.05	0	0	0
High	0.15	0.20	0.60	0.05	0	0
Medium	0.10	0.10	0.20	0.70	0.05	0
Low	0.05	0.10	0.10	0.15	0.80	0
Very low	0	0	0.05	0.10	0.15	1.0

Table 8.10: Damage to Buildings.

Type of Damage	Non-Engineered		Semi-Engineered		Engineered	
	Nos	per cent	Nos	per cent	Nos	per cent
Destroyed	312800	21	41980	24	6840	20
Very High	319782	21	33945	20	6701	21
High	389009	26	41818	24	7700	24
Medium	481673	32	56335	32	11120	35
Total	1503264	100	174378	100	31861	100
Orissa State	6677882	-	1567244	-	701649	-

After carrying out an elaborate analysis some very significant conclusions emerge. The percentage of damage suffered by the three categories of building types under various damage levels is nearly the same. Nearly 20 per cent of the buildings are destroyed, another 20 per cent are very highly damaged, nearly 25 per cent suffer high damage and 35 per cent of the buildings suffer medium damage. However the percentage of buildings damaged during the cyclone as compared to the total building stock in respective categories for the non-engineered, semi-engineered, and engineered buildings/ works out to 23 per cent , 11 per cent and 4.5 per cent , respectively. The total number of buildings destroyed or damaged as per the white paper prepared by the Orissa Government is 16,50,086. The total number of houses damaged/destroyed (by summing-up values in Table 8.10) computed is 17,09,503. The percentage error is only 3.5 per cent which is acceptable.

Contributions of a National Laboratory to Cyclone and Wind Disaster Mitigation

SERC-UNDP Project on Engineering of Structures for Cyclone Disaster Mitigation

Tests on models of buildings and structures in simulated atmospheric boundary layer flows had become an accepted design tool, and a variety of experiments were being conducted all over the world. The period during 1960 to 1980 saw establishment of large sized wind tunnels, often for special purposes. The situation existing in India was very different. Some of the aeronautical wind tunnels were using grid generated turbulent flows for model studies. The stability of turbulence was always in question. A country like India needed at least a few wind tunnels to solve the complex problems relating to wind loading and response on industrial structures such as chimneys, cooling towers, lattice towers, industrial sheds, and so on.

SERC made a project proposal to UNDP with the blessings of Government of India. The International Decade for Natural Disaster Reduction (IDNDR) also coincided with the submission of the above proposal. The fact of the matter however is that the project when it fructified, was already delayed by at least a decade in spite of continuous persuasion at various levels.

The project document was well thought out and very detailed. It was realized that in addition to creating the facilities, a few other things are also extremely important. First of these was the realization that interaction with world renowned experts was essential. The leading experts were identified and persuaded the visit the centre on consultancy assignments. The second aspect related to development of knowledge base and expertise of the scientists of SERC. Exposure to R&D programmes being undertaken by leading institutions under the guidance of leaders in the profession was the best way to achieve this. Hence a few fellowship training programmes were included. Right at the beginning it became clear that sufficient knowledge and expertise would be developed during the course of execution of the project. Hence, organisation of a number of workshops was included as part of the project proposal. These were planned such that the international experts present as consultants would take part in such workshops. Hence the programme was expected to benefit a large number of academicians, consultants, engineers drawn from public and private sector agencies who attended these workshops. The immediate objective of the project was to improve structural engineering expertise, and experimental facilities to provide safe and economical design of residential, industrial, institutional buildings; bridges, towers, and other infrastructural facilities; and to mitigate damage to structures due to action of cyclones. The major project outputs were listed as given below:

1. Wind engineering laboratory with fully equipped atmospheric boundary layer wind tunnel facility,
2. Capability attained to undertake wind tunnel testing on models of structures, field experiments on dynamic response of full-scale structures, post-cyclone

damage assessment and preparation of detailed technical reports on the above subjects,

3. Development of guidelines on construction methods for safe and economical design of structures,
4. Training of scientists,
5. Training a minimum of 100 professional engineers, architects and planners in selected technical areas, and
6. Dissemination of guidelines and development of standards for adoption in codes for cyclone resistant construction and design of structures.

Each of these activities had a bar chart and identified deliverables. Documentation was one of the hall marks of the project. All the consultants gave their reports. All the scientists who visited various institutions and conducted R&D submitted detailed reports as well. They were also able to gather lot of literature on the topic of wind engineering which were not available in public domain. At the end of the project period a UNDP-GOI joint mission evaluated the performance during the course of the project and expressed the opinion that SERC be recognized as a Centre of Excellence in Cyclone Disaster Mitigation. They felt that there is a need for wide spread transfer of knowledge and expertise generated during the course of the project to the various segments of the society. This led to another short extension to the project which was titled "Action Plan for Transfer for Technology".

Action Plan for Transfer of Technology

SERC has been closely interacting with various Government and Non-Governmental organizations and has been providing the much needed technical advice for construction of mass housing schemes in cyclone prone areas. Particularly, the centre has been interacting closely with CAPART, BMTPC, APSHCL, German and Indian Red Cross, and a number of non-governmental voluntary organizations. For successful implementation of the action plan, nodal centres which are interested in the area of cyclone disaster mitigation comprising engineering colleges, NGOs and Governmental agencies were identified covering the entire coastal regions of the country. A nodal coordinators' meet was organized to explain the action plan and to discuss the various aspects involved.

As one of the activities under the Action Plan, a total of 35 cyclone prone villages covering the coastal states of Tamil Nadu, Andhra Pradesh, and Orissa have been surveyed and the results presented in the form of a compendium. This survey is first of its kind in India, which generated a comprehensive database of the types of construction, materials and layouts of buildings in representative villages of the cyclone prone east coast and would be of immense use in preparing a vulnerability map for housing in cyclone prone regions. More than twenty special training and contact programmes were organized at several villages along the coastal states of India. More than 1500 participants, drawn from wide spectrum comprising of engineers, administrators, NGOs, NSS volunteers, mason, carpenters, panchayat members and local people, have benefited by these contact programmes for disseminating technical knowledge and expertise. These training programmes were designed at two levels, one for the engineers and the other for the artisans and the public. These evoked keen interest, and the training programmes have been extensively covered by the local media/press so that the participation by the society was greatly enhanced.

In order to disseminate the knowledge/expertise gained in the field of cyclone disasters mitigation the following book, pamphlets/posters in local languages of cyclone affected regions were prepared.

1. Brochure on Guidelines for Mitigating Damage to Dwellings due to Cyclones which were printed in English, Hindi, Tamil, Telugu, Oriya and Bengali;

2. Posters for display at Village / Panchayat Offices to be distributed by nodal centres. The titles of the posters brought out in English, Hindi, Tamil, Telugu and Oriyya are:

   ☆ Improvements to building layouts to reduce damage due to cyclones

   ☆ Improvements to roofs and walls of Buildings to reduce damage due to cyclones

   ☆ Improvements to thatched roofs and mud walls to reduce damage due to cyclones;

   ☆ Improvements to tiled/A.C. sheet roofs to reduce damage due to cyclones.

3. A book titled "Guidelines on Design and Construction of Buildings and Structures in Cyclone Prone Areas";

4. A report on Damage to Buildings and Structures due to Kakinada Cyclone on November 6, 1996; and

5. A comprehensive video cassette on R&D activities conducted at SERC, Madras, in the area of wind engineering, a typical damage survey conducted, and explanatory visual aid on cyclone resistant construction features for buildings.

In order to demonstrate to the public the effectiveness of improvements suggested in common man's dwellings to resist cyclonic forces four types of buildings with thatch roof, mud wall, tiled roof, A.C. sheet roof, RCC roof with brick wall have been constructed (Figure 8.11). These demonstration houses have been constructed at Madras, Nellore and Visakhapatnam which are some of the worst cyclone prone regions of East coast of India.

Impact of the Project

The creation of world class facilities and development of knowledge and expertise led to a number of path breaking activities being taken-up by the Centre. An exhaustive write-up on all activities is difficult, but mention may be made of the following:

A large number of wind tunnel studies on models of buildings and structures in simulated atmospheric boundary layer flows have been conducted and these include:

1. Tests on pressure models of low-rise buildings

2. Measurement of panel forces on building envelopes

Figure 8.11: Demonstration Houses for Cyclone-prone Regions.

3. Tests on chimney stacks of varying geometries with and without strakes, range of chimney heights varying from 0 m to 275 m interference studies among a group of pressure vessel structures,

4. Tests on lattice towers singly, and in a group,

5. Tests on space grid roofs,

6. Aero elastic tests on octagonal shaped umbilical tower with and without GSLV,

7. Tests to determine interference effects between chimneys,

8. Tests to study group effects on building clusters,

9. Tests to determine the mechanism of boundary layer development along the length of the wind tunnel,

10. Tests to determine pressure distribution around circular cylinders, and for evaluating correlation length,

11. Pressure measurements on models of three closely spaced building having complex geometry and whose heights were about 200 m,

12. Tests to determine the stability of bridge cross sections,

13. Response of cables with and without straps, and

14 Tests on pressure and aero elastic models of cooling towers, singly and in group.

A number of analytical investigations have been undertaken based on the data collected leading to better understanding of the phenomenon of wind loading on structures. Also specialized and unique accessories like two degree freedom model base hinge assembly for dynamic testing, rigs for sectional model tests etc., have been developed. A few technical papers published based on wind tunnel investigations are included in references.

Design of Support Towers for Wind Energy Generators

India is the third largest producer of wind energy, which is environment friendly, and renewable. The design of support towers for wind energy generators are often imported along with the machine. The performance of these have been less than satisfactory in many instances as the local conditions like soil strata, wind turbulence, etc. are not correctly included in the design. SERC, Madras, has developed a design procedure and a computer programme, which can comprehensively take into consideration the various parameters including fatigue. A number of organization have approached SERC, Madras, for carrying out the design of support towers for wind energy generators. A number of full-scale field experiments for performance evaluation of wind energy support towers have also been carried out.

Cyclone Shelters for Orissa

Cyclone shelters form as a major infrastructural facility for marooned people during the occurrence of cyclonic disturbances. Various forms for cyclone shelters have been envisaged. Realizing that excellent expertise and capabilities exist for planning and design of cyclone shelters, the Indian and German Red Cross and KFW, Germany, approached SERC, Madras to provide an appropriate design solution for the cyclone shelters. SERC designed a stilted and aerodynamically shaped cyclone shelter for use in Orissa coast with specialized foundation and raised ground level to reduce the effects of storm surge. All the existing 23 cyclone shelters in Orissa are based on the design given by SERC, Madras. The lives of nearly 46,000 people, who took refuge in these shelters, were saved during the Orissa Super Cyclone in 1999.

Workshops and Training Courses

Recognizing the quality of R&D work in the area of wind engineering many institutions like Department of Science and Technology, India, National Science Foundation, US; UNCHS, Nairobi, and ISTAD, CSIR have joined SERC, Madras, in conducting workshops / courses on wind engineering and disaster mitigation. The wind engineering group has conducted a series of workshops and courses on Wind Disaster Mitigation which attracted wide participation from administrators, planners, consultants, engineers from Public and Private Sector agencies, non-governmental organization and academic institutions (Table 8.11). The results of deliberations in Workshops and courses led to fine-tuning of the guidelines for design and construction of buildings and structures in cyclone-prone areas.

A TCDC workshop on Natural Disaster Reduction: Policy Issues and Strategies focused on much broader issues involving a holistic approach to the problem of natural disaster reduction/management. A number of international delegates from Sri Lanka, Hong Kong, Bangladesh, Thailand, Australia, and USA participated in the above TCDC Workshop. The workshop was a great success and specific recommendations on policies and strategies have been formulated. The outcome of the workshop deliberated by the august body of learned Professors, Administrators, R&D Scientists, and Voluntary organizations was expected to be useful in shaping up the National policies on Disaster Mitigation.

Table 8.11: Workshops/Courses Conducted by SERC in the Area of Wind Engineering and Cyclone Disaster Mitigation.

Sl.No.	Title	Sponsors	Period
1.	Indo-US Workshop on Wind Disaster Mitigation	DST (India), NSF (USA), SERC, Madras (India)	Dec. 1985
2.	UNCHS-SERC Workshop on Cyclone Disaster Mitigation	UNCHS (Nairobi), SERC, Madras (India)	Feb. 1986
3.	Wind Disaster Mitigation of Structures	SERC, Madras (India)	Nov. 1990
4.	Design of wind Sensitive Structures	SERC, Madras (India)	Jan. 1992
5.	Design and Construction of Cyclonic and other Extreme Wind Disaster Resistant Structures	SERC, Madras (India)	Feb. 1993
6.	Workshop on Strategies for Design and Construction of Structures to Mitigate Damage due to Cyclones	SERC, Madras (India)	Jan. 1994
7.	Engineering of Structures for Mitigating Damage due to Cyclones	SERC, Madras (India)	Jan. 1995
8.	Cyclone Resistant Design and Construction of Buildings and Structures	SERC, Madras (India)	Oct. 1995
9.	Training course on Engineering of Structures for Mitigating Damage due to Cyclones	SERC, Madras (India)	Oct. 1997
10.	TCDC Workshop on Natural Disaster Reduction: Policy Issues and Strategies	SERC, Madras (India) ISTAD, CSIR (India)	Dec. 1999
11.	Analysis and Design of Structures for Wind and Seismic Loads	SERC, Madras (India)	Feb. 2000

Conclusions

Disaster mitigation against natural hazards is complex and multi-disciplinary in nature. A holistic view on the entire spectrum of activities is needed. The efforts in the Indian context in the area of cyclone disasters mitigation are highlighted in this paper. The importance of planned R&D activities has been clearly brought out. Still there are many grey areas which require research and developmental activities. One of these pertains to full-scale instrumentation for establishing cyclonic wind characteristics along the entire coastal belt and validate the cyclonic wind speed spectrum obtained at Chennai. Also, the importance of taking the R&D results to the field has been highlighted.

References

[1]   A Global Report – Reducing Risk: A Challenge for Development" by United Nations Development Program-2004.

[2]   Abraham,A., Harikrishna,P., Gomathinayagam,S., and Lakshmanan,N. (2005), Failure Investigation of Microwave Towers during Cyclones-A Case Study, *Journal of Structural Engineering*, Vol.32, No.3, August –September, pp 147-157.

[3]   Arunachalam, S., Govindaraju, S.P., Lakshmanan, N., and Appa Rao, T.V.S.R. (2001), Across-wind Aerodynamics Parameters of Tall Chimneys with Circular Cross Sections – A New Empirical Model, Engineering of Structures, Vol .23, pp 502-520.

[4]   Chan, S.H. (1971), "A Study of the Wind Field in the Planetary Boundary Layer of a Moving Tropical Cyclone", M.S.Thesis, School of Engineering and Science, Newyork University, Newyork.

[5]   Cheung, J.C.K., Holmes, J.D., Melbourne, W.H., Lakshmanan, N., and Bowditch, P. (1997), Pressures on a 1/10 Scale Model of the Aerodynamics, 69-71, pp. 529-538.

[6]   Deb, S.K., Upadhyaya, H.C., and Sarma, O.P. (2004), "Role of Atmospheric Models in Wind Engineering", International Workshop on Wind Engineering and Sciences, WES-04, pp 142-156.

[7]   Georgion, P., Davenport, A.G., and Vickery, B.J. (1983), "Design of Wind Speeds in Regions Dominated by Cyclones", Journal of Wind Engineering and Industrial Aerodynamics, Vol.13, pp.139-152.

[8]   Goel, N.K., Kumar, S., and Gairala, A. (2004), "Extreme Wind Quantile Estimation for India using a Pooled Frequency Analysis Approach", International Workshop on Wind Engineering and Sciences, WES-04, pp 182-206.

[9]   Gomathinayagam, S., Shanmugasundaram, J., Harikrishna, P., Lakshmanan, N., and Rajasekar, C. (2000), Dynamic Response of a Lattice Tower with Antenna under Wind Loading, Journal of the Institution of Engineers (India), Vol.81.

[10]   Gomes, L., and Vickery, B.J. (1976), "On the Prediction of Tropical Cyclone Gust Speeds along the North Australian Coast", Institution of Engineers, Australia, CE Transactions, CE 18, 2, pp.40-49.

[11]   Harikrishna, P., Annadurai, A., Gomathinayagam, S., and Lakshmanan, N. (2003), Full-Scale Measurements of the Structural Response of a 50 m Guyed Mast under Wind Loading, Engineering Structures, Vol.25, pp 859-867.

[12]   Harikrishna, P., *et al.* (2002), "Site Specific Cyclone Land Fall Probability using Statistics of Historical Track Data", National Conference on Wind Engineering, pp 139-144.

[13]   Harikrishna, P., Shanmugasundaram, J., Gomathinayagam, S., and Lakshmanan, N. (1998), Experimental Studies on the Gust Response Factor of a 52 m Tall Lattice Tower under Wind Loading, Computers and Structures, Vol.70 (1/2), pp 149-160.

[14]   Lakshmanan, N. (2006), "Recent Developments on Wind Loading and Wind Environment in India", Workshop on Regional Harmonisation of Wind Loading

and Wind Environment Specifications in Asia Pacific Economies (APEC-WW 2006), New Delhi.

[15] Lakshmanan, N., and Shanmugasundaram, J. (2002), "A Model for Cyclone Damage Evaluation", Journal of Institution of Engineers (India), Vol.83, pp.173-199.

[16] Lakshmanan, N., Arunachalam, S., Seivl Rajan, S., Ramesh Babu, G., and Shanmugasundaram, J. (2002), Correlation of Aero Dynamic Pressures for Prediction of Across-Wind Response of Structures, Journal of Wind Engineering and Industrial Aerodynamics, Vol.90, pp 941-960.

[17] Lakshmanan, N., Arunachalam, S., Selvi Rajan, S., and Ramesh Babu, G. (2004), Peak factor for pressures on low-rise buildings – Background and Philosophy of a novel method, *Journal of Wind Engineering*, Vol.1, No.1, Oct., pp 19-26.

[18] Lakshmanan, N., Arunachalam, S., Selvi Rajan, S., and Ramesh Babu, G. (2004), Wind pressures on gable and hip roofed low-rise buildings in cluster, Journal of Wind Engineering, Vol.1, No.1, Oct., pp 46-52.

[19] Lakshmanan, N., Arunachalam, S., Selvi Rajan, S., and Ramesh Babu, G. (2002), Experimental Determination of Wind Loads on 200 m Tall Three Tower Blocks, Journal of Aeronautical Society of India, Vol.54, pp 335-339.

[20] Lakshmanan, N., Arunachalam, S., Selvi Rajan, S., and Shanmugasundaram, J. (2002), Determination of Peak Factor for Wind Pressures on Low-Rise Buildings, Journal of Institution of Engineers (India), Vol.83, pp 32-38.

[21] Lakshmanan, N., Selvi Rajan, S., Arunachalam, S., and Armes Babu, G. (2003), "Full Scale Uplift Tests on the Roof Element of Low-Rise Buildings", Eleventh International Conference on Wind Engineering, Texas, USA, pp 1081-1088.

[22] Lakshmanan, N., Selvi Rajan, S., *et al.* (1998), Wind Loading for Structural Design of a Long Shed Open at Gable Ends, Journal of Institution of Engineers, Civil Engineering Division, Vol.79, pp. 55-58.

[23] Lakshmanan, N., Shanmuga-sundaram, J. (1997), "Guidelines for Design and Construction of Buildings and Structures in Cyclone Prone Areas", SERC-UNDP Publication, December, 1997.

[24] Narayanan, R., Shanmuga-sundaram, J., and Lakshmanan, N., "Survey of Dwellings in Cyclone Prone Villages of Tamil Nadu, Andhra Pradesh, and Orissa", SERC-UNDP Publication, December, 1977.

[25] Prem Krishna, (2004), "Wind Disaster Mitigation Measures in India", International Workshop on Wind Engineering and Science, WES-04, pp 44-51.

[26] Shanmugasundaram, J., Arunachalam, S., Gomathinayagam, S., Lakshmanan, N., and Haikrishna, P. (2000), Cyclone Damage to Buildings and Structures – A Case Study, Journal of Wind Engg. And Industrial Aerodynamics, Vol.84, pp 369-380.

[27] Shanmugasundaram, J., Harikrishna, P., Gomathinayagam, S., and Lakshmanan, N. (1999), Wind, Terrain and Structural Damping Characteristics under Tropical Cyclone Conditions, Engineering Structures, 21, pp 1006-1014.

[28] Thepmongkorn, S. Kwok, K.C.S., and Lakshmanan, N. (1999), A two degree of freedom base hinged Aeroelastic (BHA) model for Response Predictions, Journal of Wind Engineering and Industrial Aerodynamics, Vol.83, pp 171-181.

[29] Venkateswarlu, *et al.* (1995), "Failure of Water Tower Shafts Due to Cyclonic Winds", Ninth International Conference on Wind Engineering, New Delhi, pp 1467-1478.

[30] Vickery, P.J., and Twisdale, A. (1995a), "Wind Field and Filling Models for Hurricane Wind Speeds in United States", Journal of Structural Engineering, ASCE, 121(11), pp.1691-1699.

[31] Vickery, P.J., and Twisdale, A. (1995b), "Prediction of Hurricane Wind Speeds in United States", Journal of Structural Engineering, ASCE, 121(11), pp.

[32] Vickery, P.J., Skerliji, P.F., Steekly, A.C., and Twisdale, A. (2000), "Hurricane wind Field Model for Use in Hurricane Simulations", Journal of Structural Engineering, ASCE, 126(10), pp.1203-1221.

[33] Vulnerability Atlas of India, (Earthquake, Wind Storms, and Flood Hazard Maps and Damage Risk to Housing), (1997), Report, Part:2, Expert Group on Natural Disaster Prevention, Preparedness and Mitigation having Bearing on Housing and Related Infra Structure, BMTPC, Ministry of Urban Affairs and Employment, Government of India.

[34] Wills, J.A.B., Lee, B.E., and Wyatt, T.A. (2000), "A Review of Tropical Wind Field Models", Wind and Structures, Vol.3, No.2, pp.133-142.

Chapter 9

Improving Climate Prediction Schemes with Intraseasonal Variability: A Key Tool Toward Hydrometeorological Disasters Reduction in Tropical America

Jose Daniel Pabón
Department of Geography,
National University of Colombia
E-mail: jdpabonc@gmail.com, jdpabonc@unal.edu.co

ABSTRACT

Extreme events like heavy rainfall, floods and landslides cyclically affect different regions in the world; these cycles are controlled by climate variability or the oscillations of climatological variables around normal conditions. The knowledge on climate variability is useful for prediction to support different socioeconomic processes such as prevention of disaster generated by hydrometeorological phenomena.

Today the most studied signal of climate variability is that related to the *El Nino–La Nina* – Southern Oscillation–ENSO. In spite the ENSO cycle is the most important signal of climate variability, others components of that variability induce sensible signals which generate errors in final climate outlook. One of these signals is the fluctuations with periods 30-60 days named intraseasonal climate variability.

Using data from the Colombian territory, as a special case for tropical America, an analysis of intraseasonal variability of precipitation and its role in genesis of hydrometeorological extreme events is made in this paper. The interaction between ENSO and intraseasonal components of climate variability is showed. Also, it was included some cases in which intraseasonal fluctuations not considered in prediction caused negative socioeconomic impacts and disasters.

Finally, as conclusion it is emphasized the importance of knowledge on intraseasonal variability in different regions in order to use for improving regional climate prediction schemes and reducing the risk related to hydrometeorological phenomena.

Keywords: *Intraseasonal climate variability, Extreme precipitation events, 30-60 days oscillations, Hydrometeorological disasters, Climate prediction, Madden-Julian Oscillations.*

Introduction

It is broadly known that the major percentage of disasters is generated mainly by hydrometeorological phenomena among which strong precipitation plays an important role. Extreme rainfall events produces flashfloods and landslides, hazards that in several times concrete disasters in different regions in the world. Countries have been taking several actions in order to reduce the risk associated to these hazards: improve the weather prediction, strengthen the monitoring and warning networks, introduce the risk component in planning, etc. However, these actions are still insufficient and the negative impacts continue occurring around the world.

As mentioned above, a mean to reduce the risk associated to hydrometeorological hazards is the prediction ability a community has to anticipate a given event: more time of anticipation, more possibility of risk reduction. Nowadays almost all prevention systems have as an essential tool the weather forecasting schemes operated by national weather services in a day to day basis. In spite the accurate of forecast and the warning capabilities, a day or less is not enough to develop several needed actions to reduce noticeably the risk. Then, it is necessary to explore additional means, like proposed in this paper related to the use of knowledge on climate variability and climate prediction for disaster prevention and risk reduction.

It is well known also that climate has fluctuations with different period (months (intraseasonal), years, decades), that is called climate variability (see the principles of climate variability in Hastenrath, 1996); these oscillations produce periods with warm or cold, rainy (more than normal precipitation or frequent strong rain events) or dry conditions. The knowledge on climate variability allows to develop climate prediction schemes that could anticipate with sufficient time the occurrence of special conditions with extreme events.

In the climate system many processes generate this variability. For example, the oceanic phenomena El Nino (warm condition in eastern tropical Pacific) and La Nina (cold conditions in eastern Pacific) are the cause of 2-7 years time scale oscillations of climatic variables known as ENSO cycle (Philander, 1990; Hastenrath, 1996). Additional to ENSO cycle, others signals were identified and the most relevant are a quasi-biennial component (Ropelewski *et al.*, 1992; Pabón, 1996) and the intraseasonal oscillations (Knutson and Weickman, 1987; Bantzer and Wallace, 1996; Nogués-Paegle *et al.*, 2000; Goswami and Mohan, 2001; Bond and Vecchi, 2003).

Today the most studied signal of climate variability is that related to ENSO. Currently is broadly known that extreme variations of SST in tropical Pacific generate an outstanding signal in time sequences of climatic variables in different regions of the world. El Nino and La Nina phenomena induce anomalies (drought or strong rain) in regional climate (Ropelewski and Halpert, 1987; Pabón and Montealegre, 1992; Poveda, 1994; Poveda and Mesa, 1997; IDEAM, 1997, 1998) and produce socioeconomic impacts and bring disasters to different regions. Figure 9.1 shows variability of a monthly precipitation index for several meteorological stations, located in different regions of Colombia, a sector of tropical America; for this region, *El Niño* generate precipitation below normal and *La Nina* above normal.

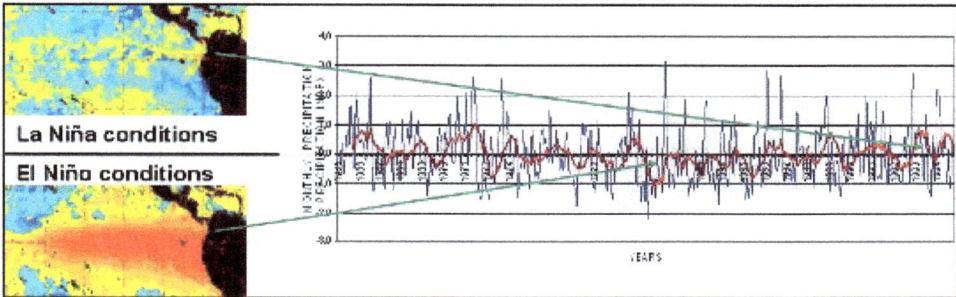

Figure 9.1: Climate Variability Represented by the Sequence of Monthly Precipitation Index for a Region in Colombian Territory, where La Nina Conditions Generate Precipitation above Normal and El Nino–Precipitation Deficit.

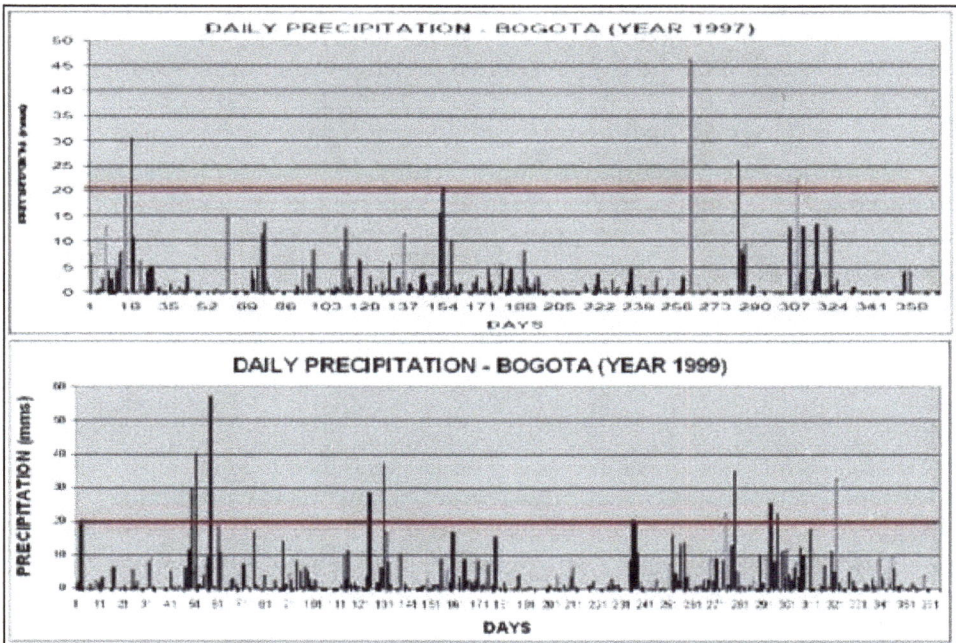

Figure 9.2: Daily Precipitation in Bogotá (Colombia) during 1997 (*El Nino* year) and 1999 (*La Nina* year). Red line indicates 20 millimetres precipitation.

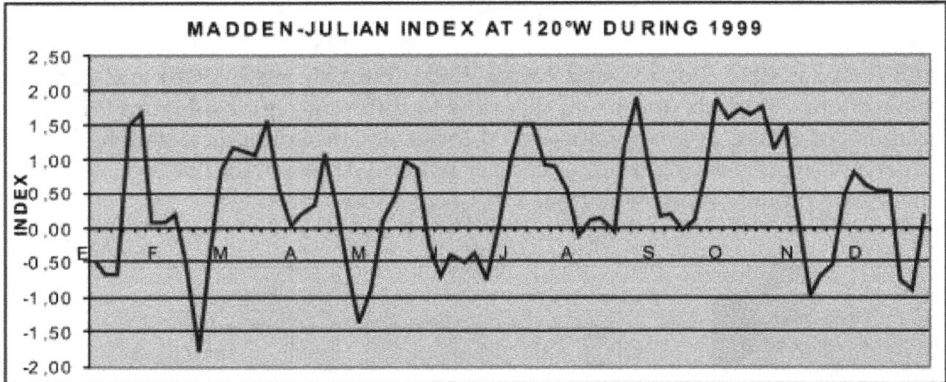

Figure 9.3: The Sequence of Madden-Julian Index (MJI) for 1999 Year (Based on data supplied by NOAA in Web page hattp://www.noaa.gov).

The effect of *El Nino* and *La Nina* phenomena also is noticeable in the frequency of extreme precipitation events. The Figure 9.2 shows the difference in the frequency of extreme precipitation events (daily precipitation more than 20 mm) during *El Nino* and *La Nina* phases of the ENSO cycle. For the region taken for this illustration, during *La Nina* more extreme rain events are observed. This feature of ENSO cycle and its relationship to extreme events is observed whole around the world and could be used to prepare with season-to-year anticipation a given region to conditions with high frequency of extreme events. In some regions this is used for medium and long range term planning for taking actions to reduce vulnerability. However, heavy rain is not occurred all time during the "rainy" period; even in rainy season extreme precipitation events activate in some periods lasting 1-2 weeks (it can be seen in Figure 9.2). Because that, prediction of the advisement of these periods is still necessary.

Heavy rain events occurred during 1999 (under *La Nina* conditions influence) produced several damages and injures in different regions of Colombia, generated mainly by flash floods and landslides (se for example data in http:// www.desinventar.org).

The outbreaks of heavy precipitation events occurred in intraseasonal scale and are mainly associated to Madden-Julian Oscillation (Madden and Julian, 1971, 1972, 1994; Hendon and Salby, 1994). A comparative analysis between Figures 9.2 and 9.3, where daily precipitation and Madden-Julian Index (MJI) for 1999 is showed, allows to confirm at a glance that there is certain relationship between these variables. The knowledge about the frequency of these outbreaks and their relationship to Madden-Julian Oscillation serve as basis to build prediction schemes that predict the presence of a high rain period with anticipation of several weeks. A first attempt to know the intraseasonal variability of the precipitation and its association with Madden-Julian Oscillation for the studied here region was made by Pabón and Dorado, 2008.

Up today a great progress has been achieved in monitoring and prediction of Madden-Julian Oscillation (Knutson and Weickmann, 1987; Mo, 2000; Nogués-Paegle *et al.*, 2000; Jones and Schemm, 2000; Jones *et al.*, 2004) and we can use MJI information displayed on the NOAA Web page as an indicator or predictor in order to anticipate periods with strong precipitation.

In this paper, an analysis of intraseasonal variability of precipitation in several regions of Colombia, northern side of South America, is done. Also, it is explored the relationship between this variability and MJO in order to use MJI as predictor in climate prediction schemes for the country.

Material and Methods

Decadal (amounts for each ten days period) precipitation data for the 1981-1990 period from meteorological station located over Colombian territory (as showed in Figure 9.4) were used for this analysis. A decadal precipitation index (DPI) was calculated using the equation:

$$PI = \frac{P - \overline{P}}{\sigma_P} \tag{1}$$

where,

PI: Decadal precipitation index

P: Decadal precipitation

$\overline{P}$: Multi-anual precipitation average for respective decade

σ_P: Standard deviation for the series of a given decade (time sequences of first decades, or second decades of the year and so on).

Results and Discussion

In Figure 9.5 the sequences of decadal precipitation index (DPI) for five analyzed points and MJI sequences for 120°W and 40°W for 1981-1990 period are presented. The comparative analysis shows that for several places in some periods DPI is in counter phase with MJI, however, in other periods this relationship changes and became in the same phase. It may be a different process still unknown for us affects this relationship.

Correlation analysis (no presented here) between MJI and DPI produced very low coefficients, however it is necessary to explore different ways to quantify the relationship that exists at least for some periods as was shown in Figure 9.5. This situation could be explained using the spectral analysis done for DPI and MJI which results are presented in Figure 9.6.

The spectrum for MJI has a strongly defined signal over frequency 0.15 cycles/ decade or 60 days period; a second signal appears over frequency 0.32-0.33 (30 days). It is easy to corroborate that DPI has an outstanding signal in the interval frequency 0.15-0.2 cycles/decade that corresponds to 50-60 days period, but this signal is not unique and there are signals over 70 days and into the high frequency interval 0.35-0.42 or around 25 days period.

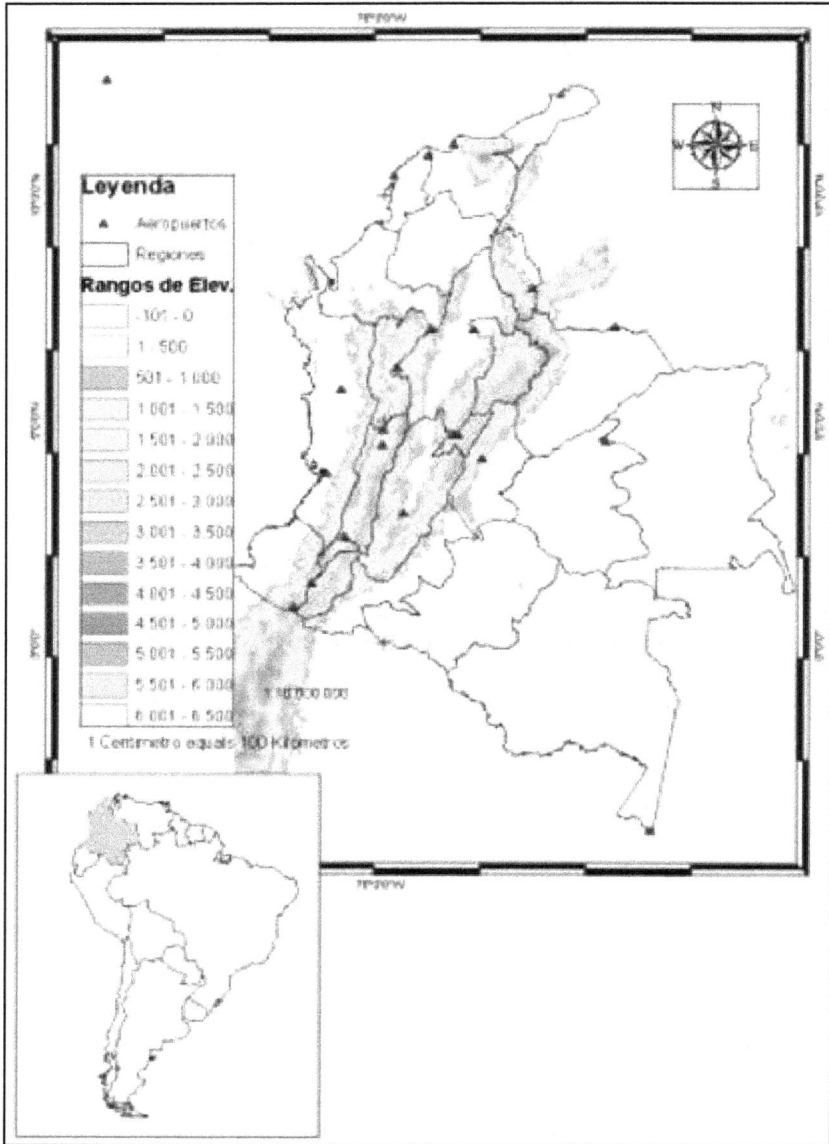

Figure 9.4: The Region in South America and Location of Meteorological Stations Used for Analysis.

The described characteristics of DPI spectra show at least two signals in intraseasonal scale which have period 20-25 and 50-70 days respectively. Precipitation of some regions has similar periodicity as MJI, in other words, has signal just around 60 or 30 days, but commonly DPI has the principal signals over 50, or 70 or 25 days period; because that in correlation analysis low values of coefficients were obtained.

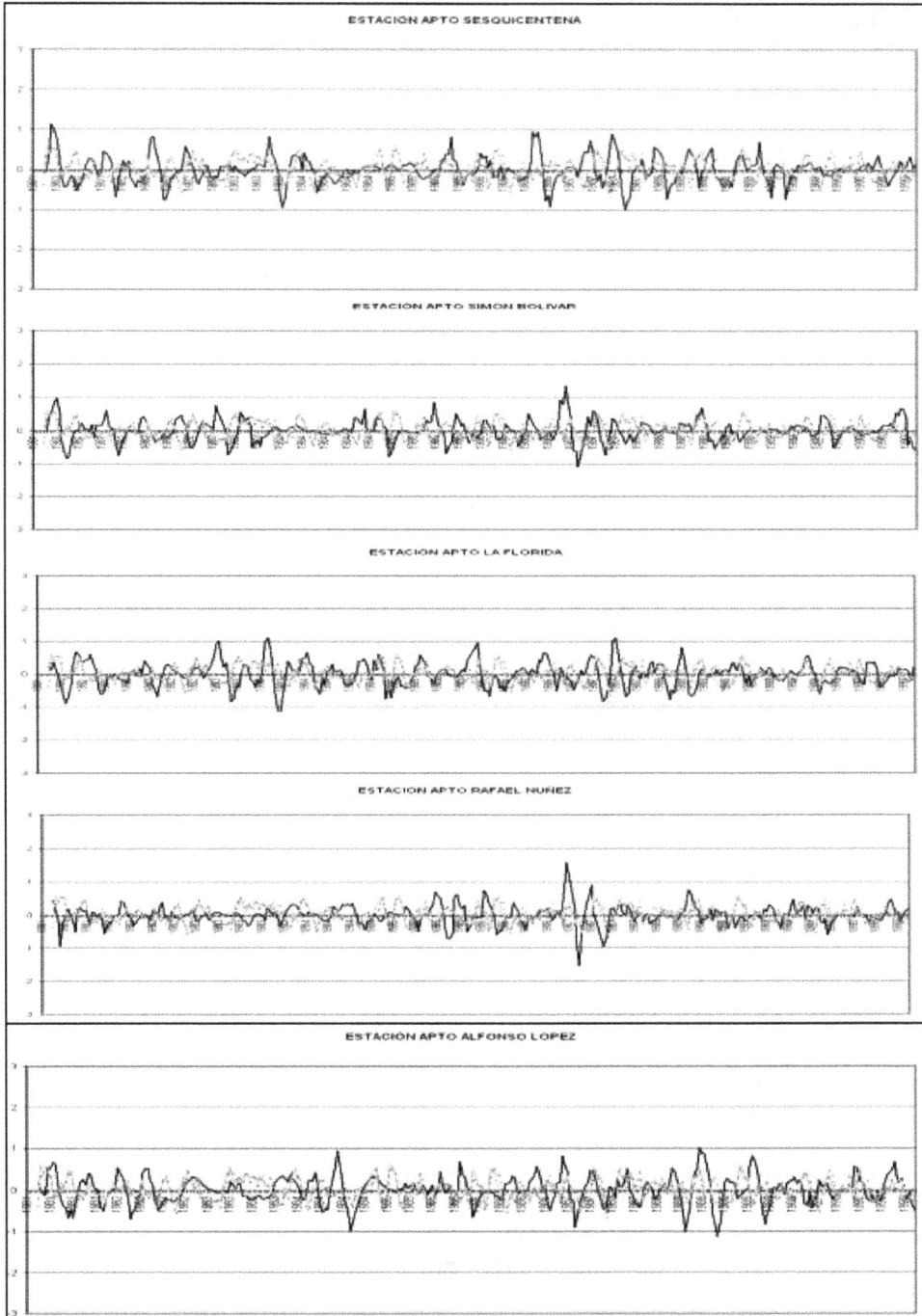

Figure 9.5: Time Series of Decadal Precipitation Index (Black line) for Five Analyzed Points with Comparison with MJI Sequences for 120°W and 40°W (Grey continued and dashed lines) in the 1981-1990 Period.

Data of MJI was taken from NOAA Web page. The components associated to 120°W and 40 l W were used. Correlation coefficients, periodogram and spectral analysis were applied to sequences of decadal precipitation index and for MJI.

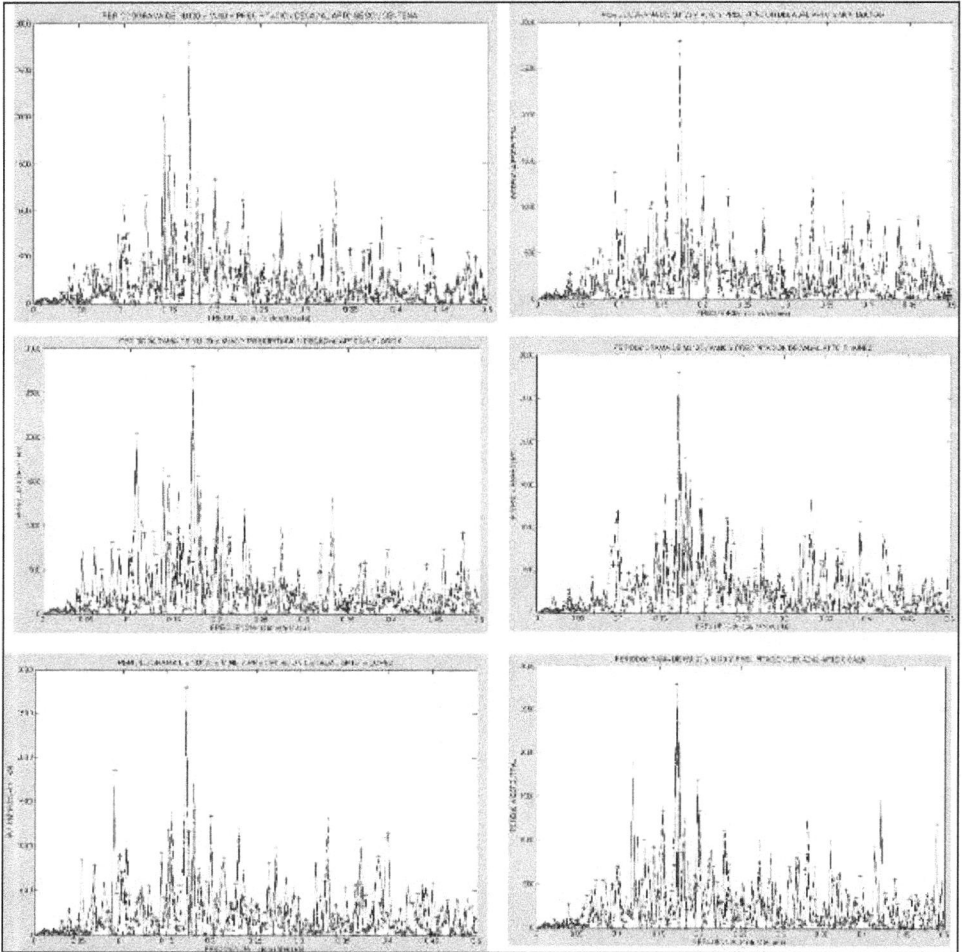

Figure 9.6: Spectral Density of DPI (Blue) and MJI (140°W red; 40°W green) Sequences from 6 (5 showed in Figure 9.5) Meteorological Stations Used in Analysis.

This analysis shows the evidences of intraseasonal fluctuations in regional (tropical America) precipitation. However, it is necessary to note that not all components of this variability are directly related to MJO, therefore, MJI could be used as predictor only for some regions. A special search is necessary to do in order to identify the regions for which MJI could serve as predictor and to identify other predictors for remain regions.

Conclusions

Phenomena like heavy precipitation events, landslides and flashfloods generate disastrous negative socioeconomic impacts in region. The analysis made above shows that the activation of strong precipitation events is related to outbreaks controlled by phases of intraseasonal climate variability. Exploring the intraseasonal variability of regional precipitation 20-25 and 50-70 days periodical components were found.

Searching the relationship between intraseasonal variability of regional precipitation and Madden-Julian Oscillation it was found that only in some regions there is a correspondence because not all regions have the intraseasonal components periods precisely over 30 or 60 days as MJI. Because that, MJI may be used as predictor only for these regions.

Acknowledgements

This work was carried out in the frame of the research project "Analysis of climate variability generated by processes different from *El Nino-La Nina*-Southern Oscillation" which is developing in the Department of Geography of National University of Colombia with financial support of COLCIENCIAS, agency which support science and technology development in Colombia (grant N° 178-2004, for 2004-2006 fiscal years). Author thanks Sandra Saenz and Jennifer Dorado who participate in data obtaining and processing, organization of time sequences, and elaboration of graphics.

References

[1]   Bantzer C H., Wallace J M., 1996: Intraseasonal Variability in Tropical Mean Temperature and Precipitation and their Relation to the Tropical 40-50 Day Oscillation. *J. of Atmos. Sc.*, v. 53, No. 21 (Nov), pp. 3032-3045.

[2]   Bond N.A., Vecchi G.A., 2003: The Influence of the Madden-Julian oscillation on precipitation in Oregon and Washington. *Wea. Forecasting*, v. 18, pp. 600-613.

[3]   Goswami, B. N., and R. S. A. Mohan, 2001: Intraseasonal oscillations and interannual variability of the Indian summer monsoon. *J.Climate*, 14, 1180–1198.

[4]   Hastenrath S.,1996: Climate Dynamics of the Tropics. Updated Edition from Climate and Circulation of the Tropics. Atmospheric Sciences Library. Kluwer Academic Publishers. Dordrecht, Netherlands, 488 p.

[5]   Hendon H.H., Salby M.L., 1994: The life-cycle of the Madden-Julian Oscillation. *J. Atmos. Sci.*, v 51, pp. 2225-2237.

[6]   IDEAM, 1997: Posibles efectos naturales y socio-económicos del fenómeno *El Nino en el* período 1997-1998 en Colombia. Santa Fe de Bogotá D.C., Julio-1997, 39 páginas + anexos.

[7]   IDEAM, 1998: Posibles efectos naturales y socio-económicos del fenómeno Frío del Pacífico (*La Nina*) en Colombia en el segundo semestre de 1998 y primer semestre de 1999. Santa Fe de Bogotá, Agosto-1998, 88 páginas + anexos.

[8] Jacobs G.A., Hurlburt H.E., Kindle J.C., Metzger E.J., Mitchell J.L., Teague W.J., Wallcraft A.J., 1995. Decade-scale trans-Pacific propagation and warming effects of an El Nino anomaly. *Nature*, v. 370: pp. 360-363.

[9] Jones, C., and Schemm J.-K. E., 2000: The influence of intraseasonal variations on medium-range weather forecasts over South America. *Mon. Wea. Rev.*, 128, 486–494.

[10] Jones C., Waliser D.E., Lau K.M., Stern W., 2004: Global Occurrence of Extreme Precipitation and Madden-Julian Oscillation: Observations and Predictability. *J. of Climate*, v. 7, pp. 4575-4589.

[11] Knutson, T. R., and K. M. Weickmann, 1987: 30–60 day atmospheric oscillations: Composite life cycles of convection and circulation anomalies. *Mon. Wea. Rev.*, 115, 1407–1436

[12] Liebmann B., H. H. Hendon, and J. D. Glick, 1994: The relationship between tropical cyclones of the western Pacific and Indian Oceans and the Madden–Julian oscillation. *J. Meteor. Soc. Japan*, 72, 401–411.

[13] Madden R.A., Julian P.R., 1971: Detection of a 40-50 day oscillation in the zonal wind in the tropical Pacific. *J. Atmos. Sci.*, v. 28, pp. 702-708.

[14] Madden R.A., Julian P.R., 1972: Description of global-scale circulation cells in the tropics with a 40-50 day. *J. Atmos. Sci.*, v. 29, pp. 1109-1123.

[15] Madden R.A., Julian P.R., 1994: Observations of the 40-50-day tropical oscillation: A review. *Mon. Weather Rev.*, v. 122, No. 5, pp. 814-837.

[16] Maloney E.D., Hartmann D.L., 2000: Modulation of eastern North Pacific hurricane activity in the Gulf of Mexico by the Madden–Julian oscillation. *Science*, 287, 2002–2004.

[17] Montealegre J.E., Pabón J.D., 2000: La variabilidad climática interanual asociada al ciclo El Nino-La Nina-Oscilación del Sur y su efecto en el patrón pluviométrico de Colombia. *Meteorología Colombiana*, No. 2, pp. 7-21.

[18] Nogués-Paegle, J., L. A. Byerle, and K. Mo, 2000: Intraseasonal modulation of South American summer precipitation. *Mon. Wea. Rev.*, 128, 837–850.

[19] Pabón J.D., 1996: Variabilidad inter-anual de la precipitación estacional en la Amazonia Colombiana. En: *"Dialogo en la Amazonia: Estructuración territorial, ética ambiental y desarrollo en Colombia.* Memorias del XIII Congreso de Geografía, 11-15 de agosto de 1994, Florencia, Caquetá). Universidad de la Amazonia–ACOGE–OEA -PROMESUP, 87-96.

[20] Pabón J.D., Dorado J., 2008: Intraseasonal variability of rainfall over Northern South America and Caribbean regions. *Earth Sc. Res. J.*, Vol. 12, No 2 (December 2008), pp. 70-88. ISSN 1794-6190

[21] Pabón J.D., Montealegre J.E., 1992: Interrelación entre el ENOS y la precipitación en el noroccidente de Suramérica. *Boletín ERFEN*, No. 31, p. 12

[22] Philander S.G.H., 1990: *El Nino, La Nina* and Southern Oscillation. Academic Press, 291p

[23] Poveda G., 1994: Cuantificación de los efectos de *El Nino* y *La Nina* sobre los caudales mensuales de los ríos colombianos. *XVI Congreso Latinoamericano de Hidráulica e Hidrología. IAHS*, Santiago, Chile.

[24] Poveda G., Mesa O.J., 1997. Feedbacks between Hydrological process in the Tropical South American and large scale Ocean–Atmosphere phenomena. *Journal of Climate*, 10, pp. 2690–2702

[25] Ropelewski C.F., Halpert M.S., 1987: Global and regional scale precipitation patterns associated with the *El Nino*/Southern Oscillation. *Mon. Wea. Rev.*, 115, pp. 1606-1626.

[26] Ropelewski C.F., Halpert M.S., X. Wang 1992: Observed Tropospheric Biennial Variability and its Relationship to the Southern Oscillation. *J. of Climate*, 5 (6), pp. 594-614.

Chapter 10

Lightning Safety Awareness Programme Especially in South Asia to Minimize Loss of Human Lives and Properties

Munir Ahmed[1] and Chandima Gomes[2]
[1]Lightning Awareness Center, and TARA-Technological Assistance for Rural Advancement,
1 Purbachal Road, Northeast Badda, Dhaka 1212, Bangladesh
E-mail: tara@citechco.net, munir_tara@yahoo.com
[2]Department of Physics, Colombo University, Sri Lanka
E-mail: gomes@phys.cmb.ac.lk

ABSTRACT

Lightning is a natural disaster and causes quite big losses to mankind. The losses include casualty of human and animal lives, buildings, communication, power system and other infrastructures. People have been helpless in overcoming the problem and remain affected. Many studies have been conducted to mitigating lightning, however, the science behind the lightning is not completely understood. Some technology has been developed in protecting the infrastructure (mainly buildings) but it has limitations.

It has been revealed that the preventive safety measures are still the main issue in fighting the lightning. Awareness among general people has been identified as the important factor. Programmes have been taken in this regard and people are working continuously. Lightning Awareness Cell (LAC) and Lightning Research Centre (LRC) have been opened in many countries and working on research and awareness building. It is quite prominent in South Asian countries and Sri Lanka, India, Bangladesh, Nepal and Bhutan have LAC. These countries have strong coordination with each other and achieved quite good success in awareness rising.

LAC-Bangladesh has been able to reach at community level with the awareness. The project encompasses the mass awareness to people by different means, technical education, demonstrations and training. Individual lightning safety awareness programme might have limitation in running along. As it is a part of natural disaster, there is a strong need for incorporating lightning as a natural disaster in National Disaster Management Programmes of each country. It should get equal importance like other disasters. Proper strategy and action plan should be adopted and implemented to sustain the management.

Keywords: Lightning, Safety, Awareness, Mitigation, Bangladesh, South Asia.

Introduction

Lightning is a natural disaster originates from the clash of clouds. It produces high voltage electricity and comes down to the earth causing damage to life and properties. Lightning is usually accompanied by thunder. The lightning disaster has been continuing from the very long past and probably from the origin of the earth. It occurs throughout the world though the severity is noticed in Africa, America and Asia.

The damage from the lightning is quite significant. Many people die due to lightning every year. According to estimation, more than 500 people die every year in South East Asia. However, based on data of Bangladesh, it appears that about 1000 people die in SAARC countries. In Sri Lanka, more than 50 people die every year. In Nepal 15, people died in August 2006 only. For an example, on the 5th April, 2002– five women working in a farm in Moneragala were killed on the spot, as the tree they were underneath for shelter from a thin drizzle, was struck by lightning.

The lives of these people could have been saved if they were given proper education in lightning protection. Not only human lives but there are losses in livestock, buildings, infrastructure, communication, power plant, tree etc. from the lightning. On the Sinhala-Tamil New year day of 2001, a thirty-year-old elephant, Seetha Menike, belong to the Dalada Maligawa was killed by lightning. The annual damage of assets and properties is very significant although there is no specific information. However, the estimated annual property damage caused by lightning in Srl Lanka reaches over US\$ 3.0 million (SARI 2006).

Lightning Protection is a peculiar topic that requires high specialization for an electrical engineer to become an expert. The field of Lightning Protection is being updated frequently at international level. New concepts, techniques and products are constantly introduced into the market and most importantly, based on the scientific discoveries, some products, techniques and even entire concepts are rejected from the standards and engineering practice.

Our engineers are at a distinct drawback of not getting an opportunity to access such knowledge. Moreover, in Sri Lanka, Bangladesh and some other countries in South Asia, does not have proper guidelines. In the absence of such guidance, the engineers are advised to follow some other recognized standards such as International Standards (IEC), British Standards (BS), Australian National Standards, American Standards (ANSI), etc. However, lightning standards will also not improve the situation unless a proper programme is not launched aiming at educating the general public and the parties concerned in Lightning Protection. Although a notable effort is made by the scientific community during the last several years to educate the public in the prevention of lightning hazards, the number of deaths, injuries and property damages due to lightning is still unacceptably high.

At present, both building and surge imported protection devices are trusted based upon the test certificates issued at foreign laboratories. The local distributor has no way of verifying the acclaimed properties of the products due to the lack of test facilities.

Almost all the products related to lightning protection available in South Asia have, therefore, been produced to withstand the characteristics of lightning currents and lightning generated fields measured in sub-tropical regions. A considerable number of scientists believe that characteristics of tropical and oceanic regions differ to what is observed in the above-mentioned regions. Thus, there has been a need to acquire more knowledge on tropical lightning. Regarding the products, there should be an independent body to check and approve lightning protection installations and import of products.

In many developed, newly industrialized as well as expected-to-be industrialized countries, a lightning localizing system has become an essential part of the day-to-day operations of the civil administration and the defence. Thus, a lightning localizing system is an essential part. In addition, the need of a lightning detection system is badly felt in meteorology, power sector, communication sector, industrial sector, defence, hydrology, ports and fisheries and the research related to lightning.

As mentioned, the lightning was a problem and causes of loss since the ancient period, not much attention was paid until the last few decades. Recently, it has got quite good attention and people across the world are working in the issue. Scientist from different part of the world is getting together. Activities are being taken for creating awareness to common people and mitigation measures are searched although it is in an early stage. Forums and workshops have been organised and ongoing.

The Lightning Awareness Centre (LAC), for example, has been established in the countries of India, Nepal, Sri Lanka, Bangladesh with the objective of awareness building among people. Bangladesh has also a Lightning Research Center at Jahangirnagar University, which works closely with LAC for mitigation measures. Several Regional Workshops were held in India and Sri Lanka with the initiative from UNESCO where participants from several countries attended. There has also been a plan for Lightning Safety Awareness up to 2010. Some programmes like South Asia Initiative for Energy (SARI) have shown interest and support to lightning programmes.

Many of the states in India and other partner countries such as Sri Lanka, Bangladesh, Bhutan and Nepal are prone to lightning hazards. Proper dissemination of lightning protection and safety measures is – therefore – expected to be able to minimize the death toll and other hazards to the human beings and live stock as well as minimize the lightning damages to properties, industries and services such as power and communication.

It has been recognized that there are two lightning seasons of the year, during which it often causes damage in the power, communication and industrial sectors and at domestic level. This amount does not take into account the billions of indirect loses due to the downtime caused by the damaged and malfunctioning equipment and lose of data in the microprocessors. The equipment failure is on the increasing trend for the last few decades due to the wide spread the use of electronics, extension of the national power grid into rural areas and the mushrooming communication towers all over the country. Many archaeological sites, which have irreplaceable high valued structures, are either totally unprotected or not-properly protected at present, and thus the structures are highly vulnerable to lightning.

Although a notable effort is made by the scientific community during the last several years to educate the public in the prevention of lightning hazards, still the number of deaths, injuries and property damages due to lightning is unacceptably high in India especially in Eastern India. According to the National Crime Record Bureau, India 1507 persons died in India during 2001 because of lightning, which is about 1 per cent of the natural and unnatural accidental deaths in the country. In Orissa (a state in Eastern India) alone, about 300 persons were struck by lightning and subsequently died in 2004. Further, the equipment damage is also increasing due to the wide spread the use of electronics, extension of the national power grid into rural areas and the mushrooming communication towers all over the country (Gomes 2006, personal communication).

In Bangladesh, more than 100 people dies and another 100 injuries every year from it. In 2006, in Bangladesh 127 people died and 78 injured and a power plant of 33 KV was destroyed and few livestock were died (TARA, 2006). There are death and property losses in other South Asian countries.

Though various lightning protection and safety measures are available internationally, but it is very limited to South Asian countries. Some engineers and scientists are involved with lightning protection but still knowledge is limited in this field. Lightning Protection is a streamlined field that requires high specialization for an electrical engineer to become an expert. New concepts, techniques and products are constantly introduced into the market and most importantly, based on the scientific discoveries, some products, techniques and even entire concepts are rejected from the standards and engineering practice.

Having limitations on safety measures equipment, there has been an emerging need for proper dissemination of lightning protection and safety measures to minimize the death toll and other hazards to the human beings and live stock as well as to minimize the lightning damages to properties, industries and services such as power and communication in South Asian countries. Although a notable effort has been made by the scientists community during the last several years to educate the public in the prevention of lightning hazards, still the number of deaths, injuries and property damages due to lightning is unacceptably high in those countries. This paper deals with the issues, achievements and requirements of lightning safety awareness programme especially in South Asia.

Basics of Lightning/Fundamentals

Lightning is a natural atmospheric phenomenon which is caused by the instability of charge distribution within a cloud. It also occurs due to the charge separation in masses of ash and dust ejected in a volcanic eruption or a nuclear explosion. Lightning physics is the study of the various aspects of lightning such as background conditions for formation, corresponding activities, variation, and distribution based on the geographical location and effects on human beings and structures.

A lightning flash originates inside a cloud, several kilometres above the ground level. Except for ball lightning which is a very rare phenomenon, lightning is simply an electric spark between a cloud and ground, between two clouds or between two parts of a cloud. The spark that jumps between the ends of two wires, which are connected to the terminals of a car battery, is a very basic form of lightning.

In the first stage of the lightning strike, a channel of charge flows towards ground from the cloud. When this channel is about 50-100 meters above, earthbound objects in the vicinity (*e.g.* trees, buildings, human beings, animals etc.) start sending upward channels of opposite charge to meet the downward channel from the cloud. One of these upward channels succeeds in meeting the downward channel first. Subsequently, a large current will flow through the object, which sent that upward channel. Then, we say that the object is lightning struck. If a given building is a tall protrusion in a certain landscape it may be the unfortunate object that sends the first upward channel that meets the downward stream of charge from the cloud.

Thus most of the life casualty takes places in the open field. The property damages occurs mainly with high-rise building and electric-linked (electricity plants, telephone lines etc.) equipments. The passing of electricity through animal bodies (especially mammals) makes huge electric shock and heat that burns and kills the animal. The lightning causes fracture to complete damage to building, bursting to eclectic transformers and equipments. In a study, *Gomes et al.* (2006), has shown that the life-causality are more (70 per cent) from indirect effect than direct strike (30 per cent) by the lightning in Sri Lanka and Bangladesh. The indirect causality occurs by the side flashing when people run indiscriminately for shelter. It usually occurs during the dry and hot period of the year. In South Asia the period of lightning is from March to June as main peak time and September-October as second peak time in many places. Figure 10.1 shows the process of lightning.

One remarkable drawback of lightning mitigation is the prejudice and misconception among general people. Most of the people are ignorant on the cause of the lightning. This is very high to the people of South Asian countries. The people tend to believe that lightning comes from the God. It is a punishment for vigorous misconduct and can't be avoided. People tend to behave badly to affected person or family instead of co-operation, sympathy, assistance and support. Some people rely on spiritual measures as a remedy.

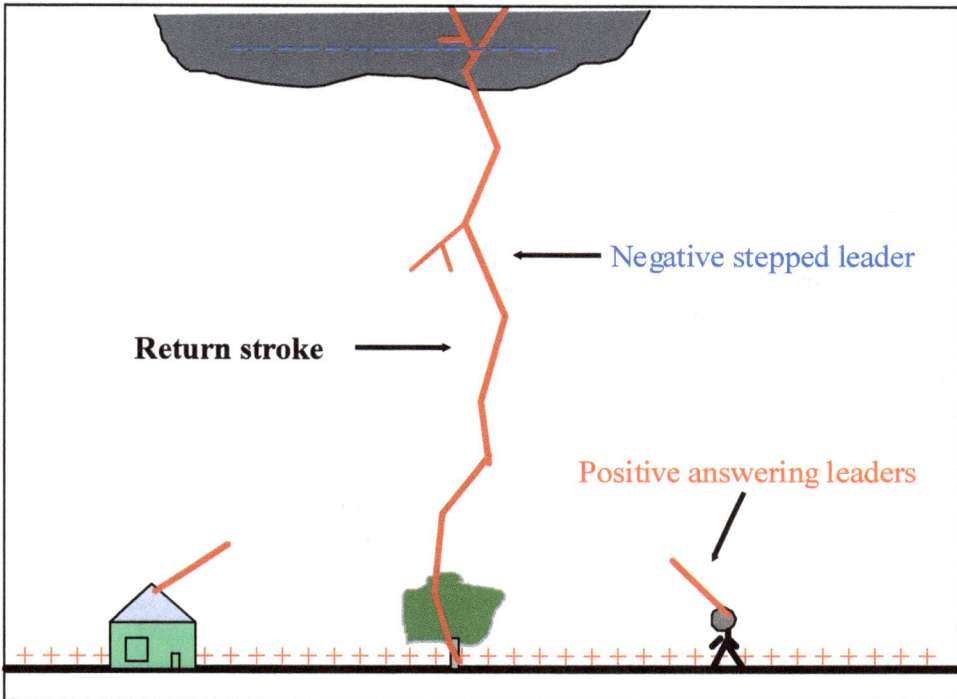

Figure 10.1: The Meeting of Negative and Positive Step Leader that Leads to Lightning Strike (Courtesy: Chandima Gomes).

Lightning Safety and Mitigation Strategy

The lightning safety and mitigation are of mainly two types: (1) Protection through technical way, and (2) prevention through taking safety/precautionary measures through building safety awareness.

Protection through Technical Ways

There has been technological advancement on the protection of lightning. The main effort has been put on the protection of valuable infrastructures. The available techniques are:

Building Protection Against Lightning

Lightning may cause damages to buildings and equipment in three ways. When a building attracts a downward lightning leader (direct strike) or attract a part of a lightning flash that hit another structure in the near proximity (side flash) it will get the maximum damage. The lightning current reaches a maximum value of about 30,000 Amperes on average but currents in the range of 300,000 Amperes are also reported. The lightning current heats its path to a temperature of about 40,000 Celsius. The enormous current involved with the lightning flash may destroy the entire power and communication networks in a building including all the equipment connected to the networks. The high temperature resulted by both the current that flows in the lines and the sparks that jump in between different parts of the building may trigger fires that will completely burn out the installation. Direct lightning (or side flashes) cause damages at a very rapid rate so that once the building is lightning struck it is very unlikely that one can prevent any of the damages to the equipment and human injuries (still one can prevent the spreading of the fire by acting quickly).

The second mode of getting lightning currents into an installation is through the service lines such as power, communication, and cable TV. Once lightning strikes a service line, fractions of the lightning current enter all the nearby buildings and may destroy all the equipments that are plugged into the system. The lightning current may also injure the users of equipments connected to service lines. This mode of intrusion of the lightning current cause less damages than a direct strike, yet service lines are subjected to lightning strikes much more frequently than buildings themselves.

The lightning current that flows from cloud-to ground is a good emitter of electromagnetic radiations. Thus once a lightning hit a nearby object (say at 500m) even the building is exposed to a strong dose of electromagnetic radiation. When this radiation passes through electronic equipments such as computers, medical equipments, military equipments etc. the sophisticated parts of them can be destroyed. This may happen even when the equipments are unplugged from service lines. However, the chances of ordinary electrical equipments getting damaged due to such radiation is very slim.

In the commercial language a building protection system is referred to a network that is meant to protect a building from direct lightning strikes and side flashes. An ordinary building protection system consists of one or several sharp pointed rods (or a mesh of metal strips/wires) installed at the roof of the building, several metal stripes or metal wires from roof to base and one or several conducting rods buried in the ground. When a downward channel comes from a cloud, the air termination sends an upward channel much faster than the other parts of the building thus the lightning is attracted to one of the rods (or to the metallic mesh). Then the lightning current is safely passed into earth through the rest of the system. Thus, instead of repelling, a lightning protection system attracts a lightning channel. And as the lightning current is safely driven into earth the building is saved from damage. The system is called the "Franklin Rod System", named after Benjamin Franklin who first proposed lightning rods in 1749.

Surge Protection

With the above-described systems a building is protected only from direct strikes and probably from side flashes. Still buildings can be penetrated by lightning currents that propagate along service lines (power, telecommunication etc.). These lines are more probable to be exposed to lightning electrical environment than individual buildings as they stretch a long distance over the land. There are special devices (surge diverters or surge suppressors), which prevent these current impulses from entering your building. A surge diverter is connected at the entrance of the service line to the installation. At high level of protection, they are connected at the power socket of electrical or electronic equipment as well. Under normal operation the surge diverter does not interfere with the line signals. In the case of a lightning invasion it provides a convenient path for the lightning current to divert into the earth without permitting it entering into the building or the equipment. Surge suppressors, which have to be connected to protect power lines, communication lines and data lines, are different from one another.

Electromagnetic Compatibility (EMC)

The sophisticated electronics in a building can also be damaged by radiation, which will be emitted from the lightning flash. Electromagnetic radiation propagates in free space. This radiation can be prevented from entering your building by taking suitable protection measures. The typical method of radiation prevention is to screen the building with a conductive material so that the building becomes a "Faraday Cage". As it is very costly and unnecessary to screen the whole building it is recommended to take safety measures only to compartments of the building where sophisticated electronics are installed (such as computer rooms, medical theatres and scanning rooms, control chambers of power plants, airports, military bases, and communication bases etc.). Design and implementation of protection systems that prevent the intrusion of undesired electromagnetic impulses through both radiation and conduction is termed Electro Magnetic Compatibility (EMC).

Lighting Prevention Through Safety/Precaution Measures (Mitigation Approach)

One of the main destruction of lightning is the loss of human lives and other living and nonliving assets. The mitigation approach is aimed at raising up the awareness of the common people.

Target Achievement

☆ *Awareness to common people*: The common people who are affected by the lightning need to aware properly with all the safety issues and precautions.

☆ *Awareness to technical people*: Technical people should be aware on the types of losses through lightning so that they can take steps in technology development for further improvement.

☆ *Research and studies*: It should be continued finding the causes and mitigation of lightning.

Way of Achievement

The awareness can be achieved through organizing workshops, symposia, seminars, posters, advertisement etc. It can be done by the more conventional way of dissemination *i.e.* workshop cum training programmes. The participants will learn on the following issues and would be useful for them.

☆ Fundamentals of Lightning system,

☆ Personal lightning safety,

☆ Common Lightning accidents,

☆ Lightning safety of household equipment safety and of outdoor workers,

☆ Alternative lightning protection system, and

☆ The usual Safety Mitigation Measures are:

Safety Mitigation Measures

Local Individual Safety Measures

In the context of Bangladesh and other similar countries in terms of physical environment farmer and fisher work in the open field and open water respectively during lightning prone time/season. The farmers work to harvest and transplant paddy during March to July during high lightning period. Same time fishing is also important as fish migrate with the onset of pre-monsoon rain. Another peak time is September-October when fishers busy in fishing soon after flood recession and farmers also work in paddy field. These two periods are prime period in Bangladesh and some other South Asian countries. Many human lives are taken away along with livestock in this period. It makes an irreparable loss to poor families that losing of the earning member or their main assets. The individual safety measures are very important and can play significant role on reducing the lightning loss.

Building Safety

The technical safety for building has been mentioned before. Other important infrastructures like fly over, bridges, airports, archaeological sites etc. are also sometimes affected by lightning and require protection. More safety measures to adopt.

Telecommunication Safety

The telecommunication (including mobile phone service providers) systems are prone to lightning. Preventive measures are required to be developed and enforced.

Electricity and Power Safety

It is another area of huge asset damage. No technical and preventive measure is in place. Action is required to save this sector as this support most of the economic activity like garments, manufacturers etc. Interruption in electricity due to lightning may not only damage equipment of that industry/office but may contribute in failing to shipment within deadline and thus consequently business loss of million dollars.

Lightning Safety Programmes Especially in South Asia

Initial Programmes

Several Regional Workshops were held in India and Sri Lanka with the initiative from UNESCO Delhi where participants from several countries attended. Physics teachers from College and University from Bangladesh also attended in those workshops. In all the workshops second author of this paper attended and worked as resource person.

One of the major recommendation of those workshop was to make more awareness programme on lightning Physics in South East Asia. Later UNESCO nominated second author (Dr. Chandima Gomes) as one of the leader in awareness education programme for this region.

Under this context and UNESCO recommendation to awaring people, scientists, academicians, Dr. Gomes implemented a project where he conducts workshop/lectures in Sri Lanka, Bangladesh and Bhutan. This project was supported partially by UNDP/UNESCO and SARI Energy.

Under this programme dissemination of Thunder/Lightning safety issues with Schools by Colombo University, Sri Lanka; for School Teachers and Others awareness workshop by TARA, for awareness programme with Scientists, Academicians by Jahangirnagar University and for Engineers in Bhutan by RBIT.

Development of Lightning Awareness and Research Centres and Programmes

To carry out programmes on lightning safety awareness and mitigation work workshops alone are not sufficient. There is a need for continuous support from institutions. As existing disaster management organizations did not keep this issue as main component, need for development of Lightning Awareness and Research Centres in the region. Thus several LAC and LRCs have been established in Bangladesh, Bhutan, India and Nepal.

Sri Lanka

The NILP (National Institution for Lightning Protection) Sri Lanka is recently formed but the executive members have done a remarkable service to the nation for the last few decades with regard to lightning protection. So far they have involved as a coordinator, moderator or active resource person with over 100 workshops, symposia conferences and training programmes at international, regional and national level. These programs have been funded/sponsored by various institutions such as UNESCO, SARI/Energy (USAID), IEE Sri Lanka Branch, Sri Lanka Institute for Development of Administration, Sri Lanka Association for Advancement of Science, University of Colombo, University of Moratuwa, Sri Lanka Telephone Regulatory Commission, Institute of Fundamental Studies etc. The executive members have pioneered in establishing Lightning related projects in Sri Lanka, Columbia, Nepal, Bangladesh and Bhutan. They also engaged with the first lightning physics research group in South Asia which has a history of over 40 years. The coordinator of the executive member panel is the group leader of South Asian Lightning Awareness Programme.

India

Kerala

The Lightning Awareness Cell, Kerala was formed in June 2005 following a dialogue initiated with the SALAP team at SARI/Energy Regional meeting held in Colombo in May 2005. The institution was formed under the umbrella of Regional Energy centre Kerala, which is a not for profit organisation registered under the Travancore Cochin Literary, Scientific and Charitable Societies Act–XII of 1955. The LAC Kerala, is now strengthened with the active contribution of a number of senior academics and professionals in the region. First committee meeting is to discuss the structure and work plan of LAC, Kerala on the 12th *July 2005*. The LAC of Kerala conducted a Lightning workshop at Mysore on 07-08-06. They also organized another programme in Cochin in March 2006.

Bangalore

Department of High Voltage Engineering under Indian Institute of Science, Bangalore has been very active in lightning related research and has designed and built the lightning protection system for the satellite launch pad at Sriharikota, India, lightning test facility for aircrafts at Bangalore, lightning protection system for HANSA aircraft etc. In addition several students have worked on lightning related area for there Ph.D. as well as master degrees at the department In addition, I have also delivered several talks on lightning at different meetings in the country.

Guwahati

The Lightning Awareness Cell, Guwahati was initiated during August 2005. The cell has been formed under the umbrella of The Energy and Resources Institute (TERI), which is a not for profit autonomous organization registered under the Society Registration Act 1856 and headquarter in New Delhi. The LAC Guwahati is now strengthened with the active participation of a number of senior academics and professionals in the region. The cell has also initiated contacts with various national and international institutes working on lightning protection and safety. LAC already organized a two day workshop on Lightning in Guwahati on 26-27 April, 2006. Planned programmes are:

☆ Workshop on lightning protection and safety for school and college teachers in various regions of the Guwahati;

☆ Awareness programmes on lightning safety for panchayat members and villagers;

☆ Awareness programme for media professionals in different regions of the country;

☆ Entrepreneurship Development Programmes involving participants from different states; and

☆ Mapping on Lightning Affected areas of India by detailed Survey.

Bangladesh

The Lightning Awareness Center (LAC), Bangladesh was first proposed and formed during the workshop on 15 December, 2004 held at CEGIS by Technological Assistance for Rural Advancement (TARA), Dhaka. Later during the workshop at Janahangirnagar University (JU) on 17 December, 2004 this was further strengthened and decided to run closely with Bangladesh Lightning Research Center (BLRC) at JU and South Asian Lightning Awareness Centre (SALAC) in Sri Lanka and National Lightning Safety Institution (NLSI), USA and other organizations. The first two workshops were conducted under South Asia Lightning Awareness Programme (SALAP) Phase-1, supported by SARI/Energy.

In recent past LAC/TARA alone and jointly with BLRC, JU and local NGOs conducted several awareness programmes/activities like workshop at CEGIS jointly with JU, workshop jointly conducted at JU, Seminar in Chandranath College, Netrokona, Bangladesh, Awareness programme with school children in Sreemangal, Bangladesh SPOSP and SERRE, Awareness programme with fisher and farmer community, workshop with NGO workers and its beneficiaries at Bhurungamari, Kurigram (TARA 2005a, 2005b, 2005c, 2005d, 2005e, JU 2005. LAC TARA also started maintaining a lightning database. In addition to work in Bangladesh LAC/TARA has also been associated with lightning safety related programs abroad. LAC representative attended SARI Energy Regional meetings in Delhi, Colombo, Dhaka and Kathamandu. It also took part in Lightning programmes in Sri Lanka, Bhutan and USA. LAC TARA also produced number of reports of its events.

Under the LAC TARAs future plan following approach and strategy has been identified:

☆ Workshop/Programmes considering most Lightning affected areas,

☆ Formal Presentation/workshop at Divisional, District and Upazila Level Programme,

☆ Rural Level Programmes in village, bazar/hats/growth centres/schools with Jarigan/Folksongs for the profession wise such as: Farmers, Fishers, Boatman, Housewives, Students, Teachers,

☆ Mobile Units with a vehicle with necessary presentation equipment,

☆ Awareness materials to develop are Posters, Brochure, Flayer, Calendar, Booklet, Bill Board, Newspaper Ad, Communicating with Principals of Colleges and other institutes to organize small presentations, communicating with Editors of local newspapers/media person to publish articles,

☆ To organize a Regional Workshop,

☆ Mapping on Lightning Affected areas by detailed Survey,

☆ New website on Lightning awareness,

☆ Making new Documentary/DVD in the Bangladesh context on lightning awareness, and

☆ Creating a database on Lightning.

Bhutan

BLARC (Bhutan Lightning Awareness and Research Center) was formed during a two days workshop held at RBIT (Royal Bhutan Institute of Technology) in Phentsholling in December 2004 with the initiation of SALAC.

Nepal

Nepal Lightning Awareness Center (NLAC) was formed in late 28 August 2006 and one workshop already conducted in Kathmandu.

Pakistan

Number of workshops and seminars conducted in Pakistan with initiation of SALAC, Sri Lanka.

Maldives and Myanmar

It has been planned to organize lightning safety programme and to initiate LACs in these two countries.

Later Stage/Present Status

Lightning Safety Awareness has been planned up to 2010. Under this plan number of awareness workshop and training are planned in Bangladesh, Bhutan, Nepal, India, Sri Lanka. Planned to have more linkages with Kenya (country experiences more than 215 thunder-days/year in some areas and Prof. Robert Jallan'go Akello has been a lightning safety advocate since the early 1980's). It is also planned to have more interaction with Lightning Data Center, St. Anthony Hospital, Colorado, USA, Lightning Research Team of Uppsala University, Sweden and with National Lightning Safety Institution (NLSI).

Conclusion

Though lightning is very age old issue, many studies have conducted so far, the science behind it is not yet fully understood. This paper does not advocate to prevent lightning but to take lightning safety measures. Many damages may be reduced if proper lightning safety measures are taken. Lightning safety measures can be achieved mainly by mass awareness, technical education, demonstrations and development of lightning and surge protection devices through research and development. It is also a need for incorporation lightning as a natural disaster and give priority in National Disaster Management Programmes.

Acknowledgements

We acknowledge the information, which is used mainly from the programme/ activities of LAC-TARA, Prof. Dr. Parashuram Sharma, Bhutan, Dr. Mannan Chowdhury, Jahangirnagar University, Dhaka, Mr. Richard Kithil, CEO, National Lightning Safety Institution (NLSI), USA, Mr. Kamal Raj, Nepal, Mr. Debajit Palit and Mr. Hari Kumar of India and Mr. Akello, Kenya.

References

[1]   JU 2005. Report on Lightning Workshop of 17th December 2004. Dept. of Physics, Jahangirnagar University, Dhaka.

[2]   SARI 2006. A Compendium of activities 2003-2006. SARI Energy, Small Grants Programme, USAID.

[3]   SARI 2006. A Compendium of activities 2003-2006. SARI Energy, Small Grants Programme, USAID.

[4]   TARA 2005a. Report on Lightning Workshop of 17th December, 2004. Dept. of Physics, Jahangirnagar University, Dhaka.

[5]   TARA 2005b. Report on Lightning Awareness workshop held at CEGIS on 15 December, 2005. TARA, Dhaka.

[6]   TARA 2005c. Report on Nertrokona Programme of 24th March, 2005; Report date 25th March, 2005.

[7]   TARA 2005d. Report on Sreemangal Programmes 06-07 May, 2005; Report Date 11th May, 2005.

[8]   TARA 2005e. Draft Report on Participation to Lightning Programmes in USA of 11-13 August, 2005; Report Date 20th September, 2005.

[9]   TARA 2006. Lightning Database. LAC-TARA, Dhaka.

Abbreviations

BLARC: Bhutan Lightning Awareness and Research Centre

BLRC: Bangladesh Lightning Research Centre

LAC: Lightning Awareness Centre

NLAC: Nepal Lightning Awareness Centre

NLS: National Lightning Safety Institution

SALAC: South Asia Lightning Awareness Centre

SALAP: South Asia Lightning Programme

Chapter 11

Capacity Development: Supporting Disaster Mitigation Activity in Indonesia

Andi Eka Sakya

Indonesian Meteorological and Geophysical Agency (Badan Meteorologi dan Geofisika),
Jln. Angkasa I No. 2, Kemayoran, Jakarta 10720, Indonesia
E-mail: sestama@bmg.go.id

ABSTRACT

Indonesian's position is very unique. Indonesia is flanked by two continents and two oceans. As one of the biggest archipelagic countries in the world, it consists of more than 17000 islands lies across approximately 6000 km width from east to the west and full of volcanic mountain often called ring of fire. That situation, added by the influence of meridional and zonal circulation, affects the complexities of the climate variation in Indonesia. Furthermore, the interaction between the Indian Ocean and the African one, so-called dipole mode, impinges on the weather at western side of the country. In the eastern part, the ENSO phenomenon, frequently induce weather anomaly especially extreme drought.

In addition to that, Indonesia lies on the three dynamic faults so-called Eurasia, Pacific and Indo-Australia plates, which move approaching each other with different speed of dilatation. This alone poses a potent and vulnerable position of Indonesia to the earthquake. Depending on the movement of the plate, because almost of the plate position is within the sea area, the sub-ducted movement often causes tsunami.

Those above condition lay the foundation of the Indonesian's vulnerability to disaster, not only because of weather or climate anomaly but also because of earthquake, tsunami as well as volcanoes. Those disasters easily originate damage and live lost, especially when people do not aware of the cause as to how disaster could happen. The awareness of the incident can lead to the early warning and can also reduce the damages as well as the victims.

This paper reports the activity encompasses the country's resource capability addresses the crucial question on disaster mitigation. It covers human resource development technologically, institutionally as well as organizationally. The ultimate objective is to discuss the steps that have been undertaken related to capacity development with regard to the disaster mitigation.

This paper outlines briefly the type of disasters that potentially occur, followed by the basic approach ion management development on the disaster mitigation. Activities that have been conducted at other institutions be it individually or in coordinated term, is also explained.

Keywords: *Disaster prone, Tsunami, Meteorology, Earthquake, Early warning, Capacity building.*

Introduction

Indonesia is an archipelagic country consisting of more than 17,000 islands that straddle the equator and face both the Indian and Pacific Oceans. Her position, which is flanked by those two oceans has, in fact, posed a crucial impact on the climate characteristics. The Pacific Ocean affects on climate characteristics more of the north – east side of the islands than of the south – west part which is dependent on the distinctiveness of the Indian Ocean.

In addition to that, the mountainous relief spans along from the north of Sumatra crossing the Java Island reaching the West Papua and stranded in the north of Sulawesi, causes a massive impact on the rain characteristic all over Indonesia. The influences among the zonal and meridional winds, characteristic of precipitation affected by the oceans as well as its position to the equatorial line, originates three types of raining in Indonesia depending on the region *i.e.* monsoon, equatorial and local. The climate abnormality, either drought or wet, can cause a hazardous dry or rainy seasons. The extreme drought often originates forest fire, whereas the rain frequently creates massive flood and land slide. It is recently notified that the transition from dry season to rainy one produces local depression sporadically and generates a local storm that ruins many houses and trees, and causes lost of life.

The potential hazard caused by many type of calamity such as local storm, air quality, earthquake, flood, forest fire, etc can cause lost of life, environmental damages as well as infrastructure destruction. This may be rooted form many items, among others due to the misunderstanding of disaster mechanism, the early warning, the management of disaster mitigation, and frequently happened is caused by panics and incapability.

From the governmental point of view, the approach to reduce the casualty involves many institutions be it governmental or non-governmental governmental. The wholeness of the approach refocuses of the mitigation management into one objective *i.e.* to safe life. It starts with the early warning system, then followed by dissemination, mobilization and evacuation of the people, and resumed with rehabilitation and reconstruction. Institutionally, this approach involves three main activities: (*i*) Technical Operation, (*ii*) Capacity Development, and (*iii*) Rehabilitation and Reconstruction.

This report discusses the capacity development undertaken by the GoI related with activity to mitigate disasters problems. The Capacity Development focuses on the effort cover the assessment of the cause of disaster and educating the people to understand the phenomenon. It includes the research and development, educating the people, developing static exhibition, campaigning through real drill on disaster as well as book publication.

Definition

UNCED (1992) defined the capacity development[1] as the activities encompass the country's human, scientific, technological, organizational, and institutional as well as resource capabilities. The fundamental goal of capacity development is to enhance the ability to evaluate and address the crucial questions related to policy choices and modes of implementation among development options, based on an understanding of environment potentials and limits and of needs perceived by the people of the country concerned.

It is therefore, based on the UNDP Briefing[2] paper, with regard to the activity took place in the Netherlands in 1991, capacity development covered: (*i*) the creation of an enabling environment with appropriate policy and legal frameworks, (*ii*) institutional development, including community participation, (*iii*) human resources development and strengthening of managerial systems. It is a long-term continuing process, in which all stakeholders participate (ministries, local authorities, non-governmental organizations and water user group, professional associations, academics and others).

The disaster relief activity involves many parties, such as local government, non-governmental organization, and central government, as well. The main objective is, to safe live as much as possible. The treatment starts exactly from the onset the disaster occurs, till the rehabilitation period.

Previously, the aid is concentrated at the post-disaster phase. Recently, realizing the powerless victim and the huge amount of resource that have to be poured in recovering process, the idea of post-disaster relief is shifted on growing the self capability on the preventing side. However, because the disaster usually covers wide area of region as well as people, thus, the effort to relief and recover the disaster impact involves many parties. The prevention endeavour starts as early as prediction of the event, dissemination of the information that makes people aware of the phenomenon, and management of disaster mitigation.

The human resource development equips individuals with the understanding, skills and access to information, knowledge and training that enables them to perform effectively. The understanding of the disaster mechanism augments their readiness. And, since the disaster especially related with nature, has a certain precursory sign. The parameters related to those sign could lead to the conclusion whether or not the potential hazard could become a disaster, and thus, necessitate the decision to

---

1   The word "building" and "development" as long as related with capacity, it will be loosely used and assumed having the same meaning.

2. UNDP Briefing Paper, *A Strategy for Water Sector Capacity Building' in Delft*, The Netherlands, 1991.

evacuation. The awareness of the people will be much help if they could be helped to access information as free and wide as possible. The knowledge of construing disaster mechanism will facilitate them to be ready as early as possible far before the disaster occurred.

The capacity development also entails organizational development. An effort that covers the elaboration of the management structures, processes and procedures, not only within organizations but also the management of relationships between the different organizations and sectors (public, private and community). In term of the system, this assists in making preparation to face the disaster, such as shelter, map of evacuation, food reserve, etc.

In addition to those, human resource development and organizational development, capacity development also needs institutional and legal framework. It is accepted that rehabilitation and reconstruction may cost a lot of money and take a lot of time. The emergency caused by the needs to fulfil the basic human need, often necessitate a quick decision for financial disbursement, which is in term of national financial administrative system may not be allowed. A legal support for this should then be made possible. Of course, accountability remains imposed.

As stated by the former Director of FEMA James Lee Witt (2002) in his book titled of "Stronger in the Broken Places" (2002): Know-how comes from knowing what matters. In a crisis, you do what you have to do–but it's better to do what you planned to do. Likewise, the capacity development effort is to cultivate a sense of readiness to face the potential disaster in a more rational, systematic and designated way.

Disaster Potentiality

Weather Abnormality

Indonesia is the world's largest archipelago, which comprises more than 17,500 islands – about 3,500 of them are inhabited. The total land area of the country is approximately 2 million square kilometers, with more than 80,000 km length of coastal line. The population of Indonesia is now over 230 million, increasing at an average of 2.3 per cent annually. Sixty per cent (60 per cent) live in rural areas (about 180 sub-districts and 42,000 villages).

Indonesian's position is very unique. As one of the biggest archipelagic country in the world, Indonesia lies across approximately 6000 km width from east to the west and full of volcanic mountains often called *ring of fire*. The mountainous topography spans along from the north of Sumatra crossing the Java Island reaching the West Papua and stranded in the north of Sulawesi, causes a massive impact on the rain characteristic all over Indonesia.

The influences among the zonal and meridional winds, characteristic of precipitation affected by the oceans as well as the topography of the precipitous land originates three types of raining occurs in Indonesia depending on the region *i.e.* monsoon, equatorial and local. The climate abnormality, either drought or wet, can cause a hazardous dry or rain seasons. Furthermore, the interaction between the Indian Ocean and the African one appears with a dipole mode phenomenon, whereas

Figure 11.1: Indonesia.

in the west part the ENSO phenomenon, frequently induce weather anomaly especially extreme drought.

The extreme drought often originates forest fire, whereas the rain frequently creates massive flood and land slide. It is recently notified that the transition from dry season to rainy one produces local depression sporadically and generates a local storm that ruins many houses and trees, and causes lost of life.

It is noted that in facing the New Year of 2007, despite the fact that it gives a bounce of hope, there were many area swept by local storm and soaked by flood. Air and marine transportation suffered from wide area of extreme bad weather and high

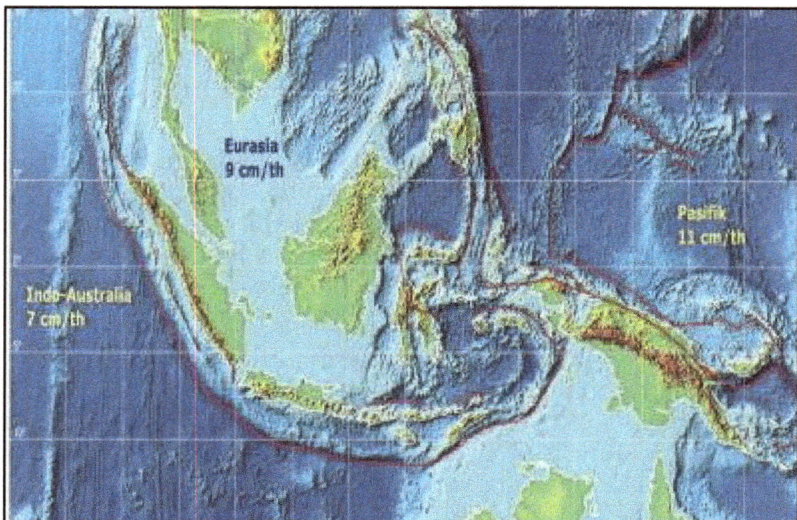

Figure 11.2: Dynamic Fault in Indonesia.

wave. This caused many fatal airplane crashed (Sulawesi, 1 January, 2007) and skidded (Ambon, Makassar), ship sunk (Tri Star, Senopati Nusantara, etc) in many places that costs of life and harbour damages.

Dynamic Plates

Indonesia also lies on the three dynamic faults so-called Eurasia, Pacific and Indo-Australia plates, which move approaching each other with different speed of dilatation. This alone poses a potent and vulnerable position of Indonesia to the earthquake. The earthquake zone spans all of Indonesia, particularly south-west Sumatra, south Jawa, Bali, Nusa Tenggara, Maluku, North Western part of Papua, Island and North Sulawesi. The so-called ring of fire which is representing the chain of volcanoes complete the potential disaster faced by Indonesian in anywhere they live. Depending on the movement of the plate, because almost of the plate position is within the sea area, the sub-ducted movement often causes tsunami.

The massive earthquake that occurred off the coast of Sumatra on December 26, 2004 was followed up with seismic activity off the island's west coast continuing into 2005, including another major earthquake in the sea west of Sumatra on March 28, 2005. During the period from 1900 to 2004, 86 tsunamis have been recorded in Indonesia, including the Indian Ocean Tsunami resulted from the Sumatra Earthquake.

Those above condition lay the foundation of the Indonesian's vulnerability to disaster not only because of weather or climate anomaly, but also because of earthquake, tsunami as well as volcanoes. Those disasters easily originate damage and live lost, especially when people do not aware of the cause as to how disaster could happen.

From Hazard to Disaster

A phenomenon that causes potentiality of damages, of live lost, or of environmental destruction can be categorized as a hazard. Whereas, disaster is an event or chain of events resulted by nature or human activities or both, causes casualty in the form of human hardships, lost of wealth, environmental damage, infrastructure and public utility destruction and disrupting people/society daily live. People rarely achieve a certain level of readiness and respond on the impact of hazard due to geographical, societal, economical, political, cultural as well as technological condition within a certain area and a period of time. The degree of vulnerability describes that level of readinesm, so that can react at the proper time to face the calamity. In the condition of facing the potential hazard, there always is available opportunity of losses. The so called risk may form on life lost, injuries, illness, threat, insecurity, evacuation, damage or lost of wealth, disturbance on conducting routine activity, in a certain area or period of time caused by hazard or disaster.

As explained previously, there are many type of disaster potentially occur in Indonesia, ranges from local storm, extreme drought, forest fire, flood, land slide, volcanic eruption, earthquake and tsunami, to name a few. This potential hazards will become worse and it may cause disasters. There are many reasons which cause a potential hazard becoming a disaster. The lack of understanding on the characteristics of hazards/disaster is, among others, common. The exposure of nature on a certain condition cannot be understood as a sign of the worse one. The national development often touches upon the issue of equilibrium exploration on nature. The acceleration of development cannot be run after the self resurgence of the nature to replace the missing balance. This degrades the level of support of the nature resources and increasing the vulnerability. An information gap can even also be one aspect that give a reason why a potential exposure of nature becoming a disaster. The lack of information adds to the powerless capability in facing the danger. Furthermore, the non-existing early warning systems, be in the form of information publicly published in book, brochure, flyers as well as hard/soft-wares, create a condition of surprises when disaster struck. The condition that none-of-the people could be expected to react in a rationale way. The situation becomes worse as people run panicky to flee the danger.

The improvement of knowledge, especially in the mechanism and mitigation of disaster, will facilitate people to increase their readiness and level of confidence. Research on science and technology shares the effort of fulfilling the needs. The implementation of effort will increase the level of readiness and confidence on facing the potentiality of hazards.

The exertion to manage disaster mitigation expands not only being focused on the rehabilitation and reconstruction or post-disaster, but also includes the endeavour during pre-disaster or prevention. This attempt covers a cycle of disaster management that consists of pre-, during and post disasters. The prevention efforts take into account to improve the readiness of the people, the mitigation as well as the establishing early warning systems. During the disaster period, an emergency treatment involving various kind of institution is given to the victim this include search and rescue, evacuation as well as mobilization of aids. Rehabilitation and reconstruction are commonly initiated after the disaster.

From the governmental point of view, the approach to reduce the casualty involves many institutions be it governmental or non-governmental governmental. The wholeness of the approach refocuses of the mitigation management into one objective *i.e.* to safe life. It starts with the early warning system, then followed by dissemination, mobilization and evacuation of the people, and resumed with rehabilitation and reconstruction. Institutionally, this approach involves three main activities: (*i*) Technical Operation, (*ii*) Capacity Development and (*iii*) Rehabilitation and Reconstruction.

It has been mentioned that capacity development – which is reported in this paper – concentrates on human resource development, research and laboratory establishment. Result on the assessment of disaster mechanism, or implementation of research results, or new methodology of improving people readiness will be used by both operation to establish the early warning and the institution responsible for rehabilitation and reconstruction.

Activities of Capacity Development

Virtually, every year, flooding occurs in several locations throughout Indonesia during the rainy season, while droughts occur during the dry season and landslides occur in various places. Indonesia has 128 active volcanoes, 70 of which present a threat, and 500 dormant volcanoes. Because the soil in the volcanic zones is fertile and good for agriculture, there are high population densities in these areas, increasing the risk. There is also great potential for forest fires in Indonesia, as the nation boasts the second largest tropical rain forest in the world. Forest fires cause not only environmental degradation, but also air pollution, and thus have a significant impact on people's daily lives. Forest fires that occurred in 1997 and 1998 caused smoke pollution in several other Asian countries in addition to Indonesia. At present, it seems that forest fire become a regular occurrence when entering the dry season.

In 2006, especially, flood occurs in many places. Entering the rainy season on December, Aceh smashed up by flood that damaged 49 villages. Approximately 200,000 people were evacuated, ruining 14 bridges and about 42,000 houses damaged. The big flood has isolated almost all the area, road remained impassable and people starved due to bad weather that constrained relief aid. The bad weather affect marine and air transportation. Entering the New Year, at least 5 ships noted sunk in various regions in Indonesia due to high ocean wave. Many people died. Right in the eve of the New Year, one plane missed in the area of Sulawesi, and still unknown until this paper is written. More than 100 passenger's fate is still unheard of.

On the other hand, Indonesia lays on three dynamic plate so-called Eurasia, Indo-Australia and Pacific plates which move approaching each others. This poses an additional potential disaster in the form of earthquake. The earthquake zone spans all of Indonesia, particularly south-west Sumatra, south Jawa, Bali, Nusa Tenggara, Maluku, North Western part of Papua, Island and North Sulawesi. The so-called ring of fire, which is representing the chain of volcanoes complete the potential disaster faced by Indonesian in anywhere they live. The massive earthquake that occurred off the coast of Sumatra on December 26, 2004 was followed up with seismic activity off the island's west coast continuing into 2005, including another major earthquake in the sea west of Sumatra on March 28, 2005. During the period from 1900 to 2004, 86 tsunamis have been recorded in Indonesia, including the Indian Ocean Tsunami resulted from the Sumatra Earthquake.

Concerned about the impact of disaster, especially after Aceh's Tsunami, the government initiated a more coordinative approach. In the capacity development side, the State Minister for Research and Technology (SMRT) is appointed as the coordinator. It involves University, Research Institutions as well as Department. The overall approach of the capacity development can be divided into two large groups of activity: structural and cultural ones.

Structural approach is aimed at improving the support of methodology in term of technology, whereas the cultural one is intended to upgrade people's awareness and readiness. Research and Development activity, dissemination of knowledge pertaining to early warning system through workshop and training, among others, belong to the structural part. On the cultural side, the training and workshop facilitated with an intention to facilitate as much as possible for non-experts, ordinary people and society. It is, therefore, the participant is required more from society. The arrangement is much looser than the training held for expert and officers. In Indonesia, these trainings had been held in various places. The local government, concerned with their people, initiated these training by inviting experts from various agencies such as BMG for early warning, LIPI for evacuation methodology, BAKORNAS for readiness.

Tsunami simulation drills have been organized in the last two years commemorating the Aceh's Tsunami in Padang on 26 December 2005 and in Bali on 23 December 2006. The first drill started with the communication between BMG-HQ in Jakarta with the Mayor in Padang informing the big scale earthquake. The DG of BMG pressed on the alarm button that directly switched the early warning alarm on in Padang. The Mayor commanded all authority bodies to start leading the evacuation of the people following the previously prepared evacuation map. The drill had been supported by Local Government of Padang – West Sumatra, Domestic Affair Ministry, SMRT, LIPI, BMG and Non-Government Organization.

Likewise, in Bali, SMRT was the focal body coordinating the drill. Rather different with the drill in Padang, the Tsunami simulation in Bali involved the international body such as PTWC, JMA as well as UNESCO. In addition to the drill in Bali, the second commemoration of Tsunami was also hold in Padang where the President pressed the newly built siren as part of the instrument of early warning system.

Furthermore, several books had also been published containing information about Tsunami and the way how to cope with when it happens. Signboards to direct the evacuation have been put in place in several places in Padang and Bali. Moreover, the Department of Culture and Tourism has even established a shelter for evacuation in Bali.

On the early warning parts, BMG supported by LIPI, BAKOSURTANAL, Ministry of Information and Communication developed National Tsunami Early Warning System (NTEWS). Grants from other countries such as Germany, China, Japan and other have been received. It is used to develop the system as well as to complete the seismometer and accelerometers. Thanks to the Government of Germany that granted a lot in this regards. One of the activities was arranging a workshop on Earthquake Information and Tsunami Warning Centre in March. The workshop was participated by many agencies that involved in establishing NTEWS. Figure 11.3 shows the workshop's participants. Furthermore, an expert has been also attached for a longer time to establish the system.

The research and development activity is supported by SMRT funded partly through the research incentive scheme. At BMG, research and development activity is conducted within one of the organizational unit, where some result has at present being used internally. Figure 11.4 depicts the number of research won the incentive research fund from SMRT. Some of the previous results have been utilized to support daily operational activity at BMG, like Tsunami Numerical Simulation, Rainy Season Prediction Zone and Fire Weather Index.

The Rainy Season Prediction or Climate Zone is usually distributed to the departments and agencies related to BMG, such as Agriculture, Forest and Public Works. Using the conventional method combined with the numerical one, BMG could

Figure 11.3: Participant of the Workshop on Earthquake and Tsunami Warning Centre.

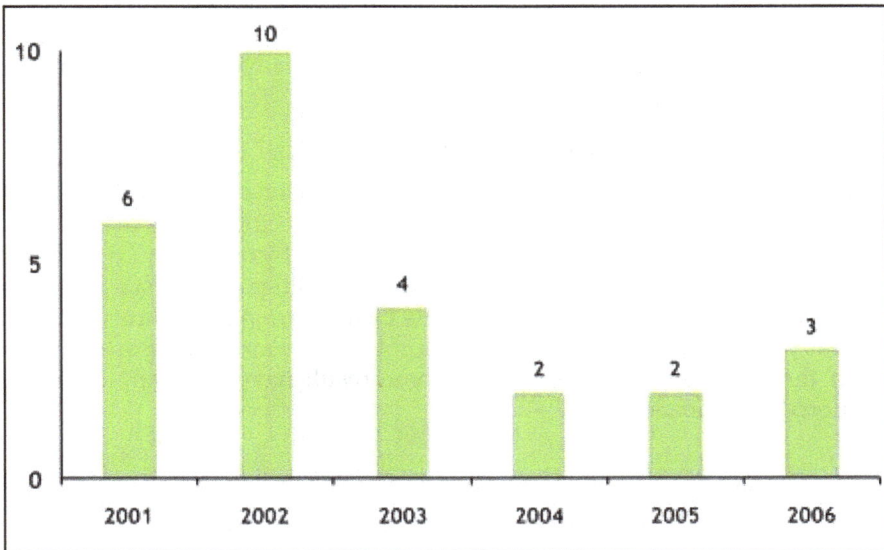

Figure 11.4: Number of Research Funded through the Incentive Scheme run by SMRT.

proof that there are more than 200 variation of climate on set in Indonesia. The information is very useful for farmer, so that they can decide to plant in the right time.

At BMG, measuring data, such as precipitation, humidity, temperature, wind speed and direction, is part of daily observation. Combining those parameters with information concerning the fine moisture, information concerning the Fine Fuel Moisture

Figure 11.5: FDRS for 4 – 5 December, 2006.

(FFMC), Duff Moisture (DMC) and Drought Codes can be obtained. Further development of those could be used to develop Fire Weather Index (FWI). FWI describes the level of potentiality of a region prone to fire due to the drought level. The result of so-called Fire Danger Rating System (FDRS) is used daily as part of the prediction pertaining to forest fire.

Concluding Remarks

Indonesia's position has been outlined in term of climate as well as geographical point of view. Indonesia is located in the area potential to any form of disasters caused by nature (earthquake, climate and weather) as well as by human (air quality, urban, etc). The capacity development related with disaster mitigation has started right after Aceh's Tsunami in the early 2005. The activity includes training, awareness campaign as well as publication. Apart from a little number of research on disaster mitigation funded through the incentive scheme, some result have been applied for daily operational activities.

Chapter 12

The Disaster Experience and Disaster Managment System of Brunei Darussalam

Pg Sabli Bin Pg Damit[1] and Sallehuddin Bin Haji Ibrahim[2]
National Disaster Management Centre, Brunei Darussalam
E-mail: [1] *pgsabli_ndmc@brunet.bn,*
[2] *sallehuddin_ndmc@brunet.bn*

ABSTRACT

Brunei Darussalam geographically lies on the northern-west of Borneo island facing the South China Sea and located between the belt of the meteorological disasters such as drought which caused forest and bush fires, regional seasonal haze, seasonal wet monsoon which cause floods and land slides, prevailing monsoon season that might caused strong winds.

The most severe disaster event that caused a great loss of million of Brunei dollars was the *El Nino* phenomena of 1998 that the nation had overcome. With such disaster, it was the first time in the wider scope that various ministries, government agencies and non government organizations, private sectors, villagers and individuals working together hand-in-hand in fighting for the extinguishment of the large scale of forest and bush fires which was never been encountered from the previous history of Brunei Darussalam. Followed by the *La Nina* phenomena of 1999 the year after the longer period of devastating episode, once again the nation have encountered a tremendous and unpredictable circumstances of down pour which causes total flooding in the predicted prone areas and flash flood on the populated lowland areas and damages a lot of infrastructures and affected on the public daily activities.

With such experiences, disaster management is very much an on-going national requirement, important to the Government and people alike. As the nation develops, disaster management has a special significance because of the increasing dangers posed by such hazards having an effect on our oil and gas based economy. The simple is that the more assets we build up, the more we stand to lose in the relationship between contemporary disaster threat and the losses it may impose. It follows that any action which can be taken to reduce disaster related loss must be seen as logical and desirable in cost benefit terms.

The establishment of The National Disaster Management Centre enacted under the Disaster Management Orders 2006 has a clear definition of the national disaster management policy to deal with all aspects of disaster threat. The main areas covered by this strong legislation are establishment of the National Disaster Council, National Disaster Management Centre, National Disaster Management Plan, Disaster Situation, Powers, Offences, Compensation and Regulations.

Keywords: *Meteorological, Disaster belt, El Nino, La Nina, National disaster management order, Regional issues.*

Introduction

Brunei Darussalam (Figure 12.1) lies on the Northern-west of the island of Borneo, 443 km north of equator. Its immediate neighbour is Malaysia state of Sarawak and Sabah. The northern part consists of a coastline of about 161 meters facing South China Sea. The Land area is 5,765 sq.km. the topography consist of low hills, dense forests and mangrove swamps. The Population is about 348,800.

Located within the meteorological disaster belt, is conscious of the geography of disaster and must be prepared to deal with the situation in relation to regional and

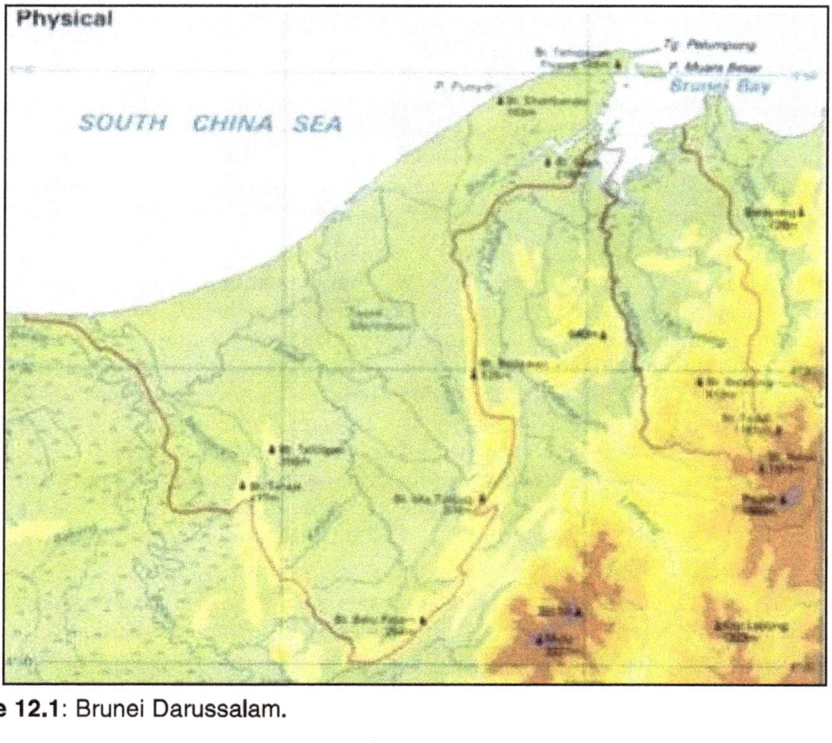

Figure 12.1: Brunei Darussalam.

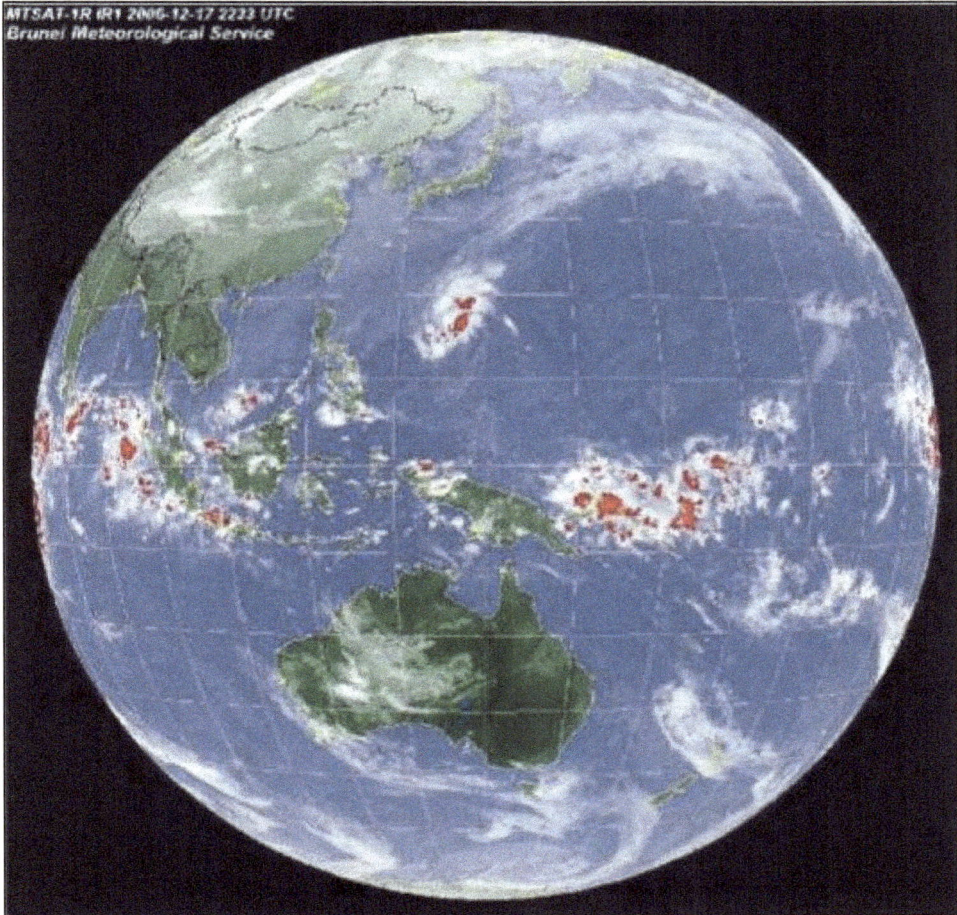

Figure 12.2: Location of Brunei Darussalam within the Global Meteorological Disaster Belt as Shown through the Satellite Image.

local circumstances. The disaster, be it natural or technological causes enormous destruction and human suffering. Disaster management is very much an on-going national requirement, important to His Majesty's Government and people alike.

As the nation develops, disaster management has a special significance because of the increasing dangers posed by such hazards having an effect on our oil and gas based economy. The simple is that the more assets we build up, the more we stand to lose in the relationship between contemporary disaster threat and the losses it may impose. It follows that any action which can be taken to reduce disaster related loss must be seen as logical and desirable in cost benefit ratio.

Figure 12.3: Thunder Storm in South China Sea, Coast Line of Brunei.

Actual Past Significant Disaster Incidents in Brunei Darussalam

The natural disasters that Brunei Darussalam has always confronted with were mostly due to the meteorological disaster. The types of such disaster were consisting of forest/bush fires, haze, flood, strong wind, land slides and others. Long before the establishment of National Disaster Management Centre, the emergency plan of the respective ministry and departments having statutory responsibility was put to operation for the following disaster incidents:

Natural Disasters

1. The Haze situation over Brunei Darussalam during the *El Nino* episode of 1998;
2. The annual haze pollution due to forest fire in neighbouring countries;
3. The flood situation during *La Nina* episode of 1999;
4. Land slide;
5. Strong winds.

Technological Disasters

1. Water Village fire;
2. The 'Big John' oil barge fire at sea;
3. The Rasau oil blow-up in Belait District.

The Experience of the Haze Situation Over Brunei Darussalam during *El-Nino* Episode of 1998

The Climate Change

The *El Nino* phenomenon (Figures 12.4 and 12.5) struck Brunei Darussalam began in early January 1998 and ended in April 1998. The land mostly from the natural vegetation of the rain forest and commercial vegetations were badly damaged

Figure 12.4: The Aerial View of Forest Fire.   **Figure 12.5**: The Smoke that Create Haze.

and was estimated of about 6,200 hectares were burnt down and estimated of loss of B$ 1.65 million dollars. At that time, the Department of Civil Aviation through its Meteorological Unit has stated that the average rainfall decreases and the temperature increases thus supporting the spread of forest and bush fires through out the whole of Brunei Darussalam. (Graphs 12.1 and 12.2).

The Impact on Environment

The deterioration of air quality (Figure 12.4) has impact on communities health, such as respiratory and eye problems. Besides that it was also disrupted on people's daily activities where at certain levels of extreme alert, the schools have to be closed down for several days. The Civil Aviation Department has to take extra precaution on its daily aircraft activities.

Some of the affected main roads have to be closed and the passage users were diverted by the Police to other alternative passageway in order to avoid further destructive traffic accidents due to the poor visibility which may cause destruction and fatalities.

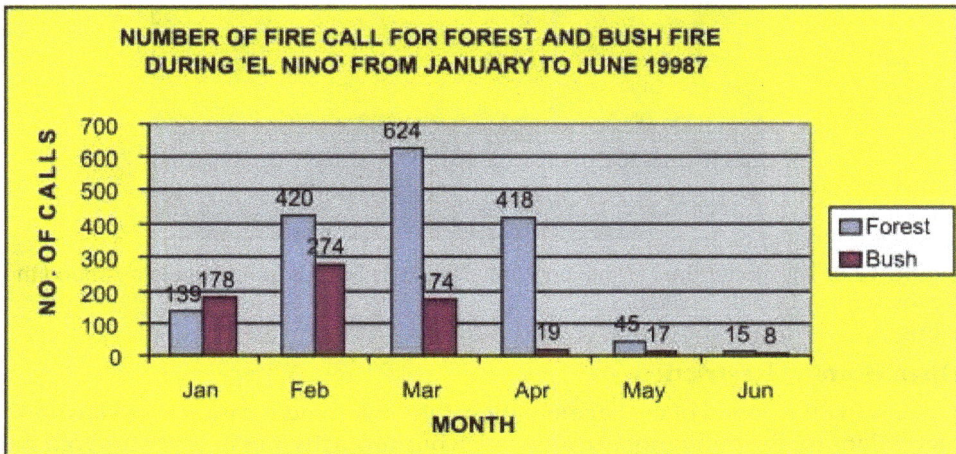

Graph 12.1: Number of Fire Calls for Forest and Bush Fire.

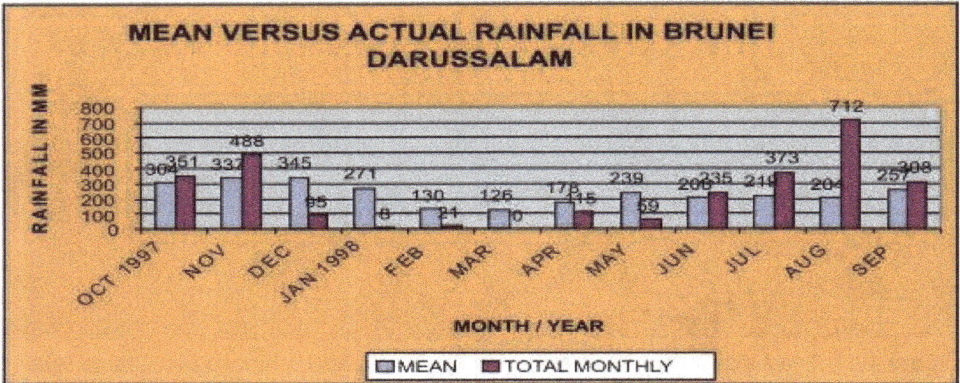

Graph 12.2: Mean versus Actual Rainfall in Brunei Darussalam.

Figure 12.6: The South East Monsoon Wind Direct the Smoke to Accumulate Around the Northern Part of Borneo Island.

Open Burning Restriction

During the period of the haze situation, the Government has imposed laws and regulations for the public not to do any open burning activities in order to avoid the additional emission of smoke which will worsen the haze situation. The Police personnel has to take patrol and take action for any person who disobey the regulation being imposed or otherwise if found guilty of such offence, the penalties will worth of B$100,000.00.

Though Brunei Darussalam is a very small country, the smoke emission from the forests and bush fires all over the state, also pay a tribute to the exports of the haze to the region.

The Lowland Peat Swamp Forests

One of the major problems during the dry seasons which create and prolong the duration of burning is the types of vegetation that mostly the Borneo Island have. It isn't only the thick forests that burnt but the peat swamp forest, in context with Brunei Darussalam lowland areas, was inevitably in a vast scale and due to its characteristics as a good fuel for combustion will burn actively especially during the dry season. From layers of woody debris too waterlogged to fully decompose slowly deposited over thousands of years, the carbon-rich peat strata have been known to reach a thickness of up to 20 meters.

Fire Suppressions

Underneath the ground of the peat swamp once affected by fire, it will leave behind a hollow or empty space which hindered the fire personnel to extinguish the fire and manoeuvres the portable equipment by means of conventional tactics on the uneven land surface (Figure 12.5) With inadequate supply of water from the open sources, the mission would not be possible unless with the application of the heavy machinery such as the excavator to dig an artificial dam (Figure 12.6) at different angles of the fire locations so as for the storage of sufficient water resources for the needs of the next fire extinguishment operation so as to fulfil the 'zero smoke emission' mission. The bull dozers will also be used to create a path along side or through the affected forest areas for the purpose of fire break so as to avoid any further spread of fire to the unaffected areas (Figures 12.7–12.10).

Air Surveillance

The formation of 'Airborne Command' based on 2nd Squadron of Royal Brunei Air Forces has become a great advantage during the haze operation. With the motto 'clear view operation' the aircrafts belong to the Armed Forces and the rented helicopters from Erickson Helintanker of United States of America has strategies in putting out the forest fire that was inaccessible by any means of land operation equipments. Through the aerial water jet and sweeping techniques from the Helintankers and water bombing from the fire buckets make the situation of fire spread be under controlled. For this purposes the Government have spent million of dollars in order to safe the environment and the public health and safety (Figures 12.11–12.15).

Figures 12.14 and 12.15 shows that the helicopters from Royal Brunei Armed Forces getting buckets full of water from open sources and bombard directly to the scene of fire.

Figure 12.7: The Fire Personnel Extinguish the Peat Swamp Forest by Total Flooding.

Figure 12.8: The Artificial Dam Used for the Storage of Water Resources.

Figure 12.9: The Excavator Machinery Used to Clear the Trees for Isolation.

Figure 12.10: The Path of the Fire Breaks being Cleared by the Excavator.

Figure 12.11–12.15

Haze Technical Task Force

The Government has established a National Disaster Committee on Haze 1998 which consist of various ministries such as Prime Minister Office, Ministry of Development, Ministry of Home affairs, Ministry of Health, Ministry of Defence and Ministry of Education.

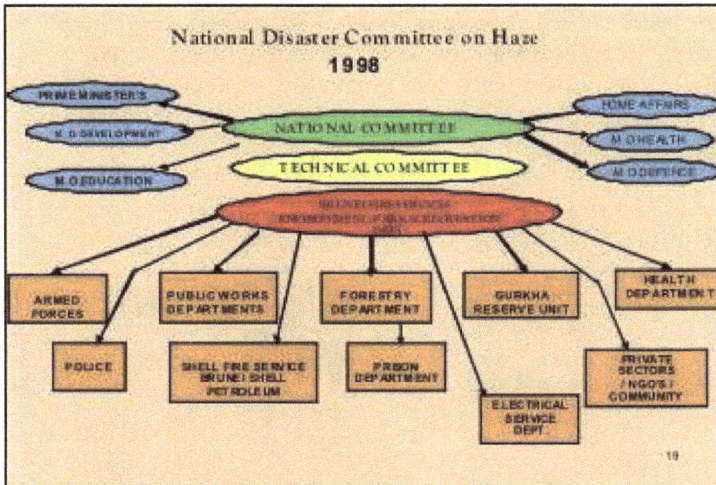

Figure 12.16: National Disaster Committee on Haze, 1998.

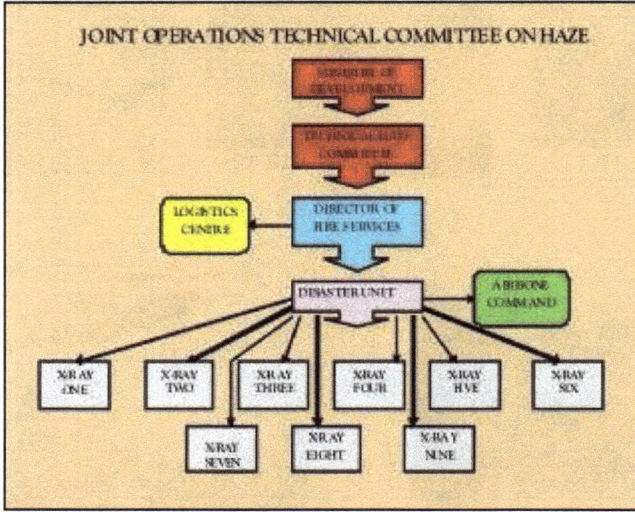

Figure 12.17: Joint Operations Technical Committee.

Joint Operation Technical Task Force

The Ministry of Developments plays important roles as the decision maker in the Joint Operation Technical Committee. Whereas The Technical committee were consists of the representatives from the various agencies dealing with the fire suppression. The Department of Fire and Rescue take the lead as the overall in-charge of the operation (Figure 12.17).

Community Involvement

During the haze period, the community especially from non government organization, public sectors, public and individuals pay tribute to the nations by assisting the technical task force and the front liners probably in giving moral support by means of sponsoring refreshments, whereas machinery and other equipments for the purpose of extinguishing the fire. Not to forget that the head of villagers and the people also pay an important role in giving hand during the operation.

The Experience of Flood Situation during *La Nina* Episode of 1999

Figures 12.18 shows the bridge, retaining wall and vegetation along the river bank were ruined due to the extra ordinary heavy down pour thus exceeding the

Figure 12.18: The Ruined Bridge due to the Extra Ordinary Heavy Down Pour.

capacity of the river located in the capital of Bandar Seri Begawan, Brunei Darussalam

The Climate Change

The *La Nina* phenomena struck Brunei Darussalam in early January 1999 and ended in April 1999. The rivers on the mainland especially in rural and lowland areas were badly damaged due to the heavy down pour and estimated a large amount of losses. Department of Civil Aviation through its Meteorological Unit has stated that the average rainfall increases through out the whole of Brunei Darussalam during the period of *La Nina* episode.

The Impact on Environment

As the average rainfall increases and the water level raised, a moderate flood, flash flood struck on some of the lowland urban and rural areas and landslides event on the hilly areas. Some of the main roads were affected and these events reflect the difficulty to the community's daily activities as for the community transportation breakdown for several days. Some schools were closed temporarily; fisherman and agriculture activities were also affected.

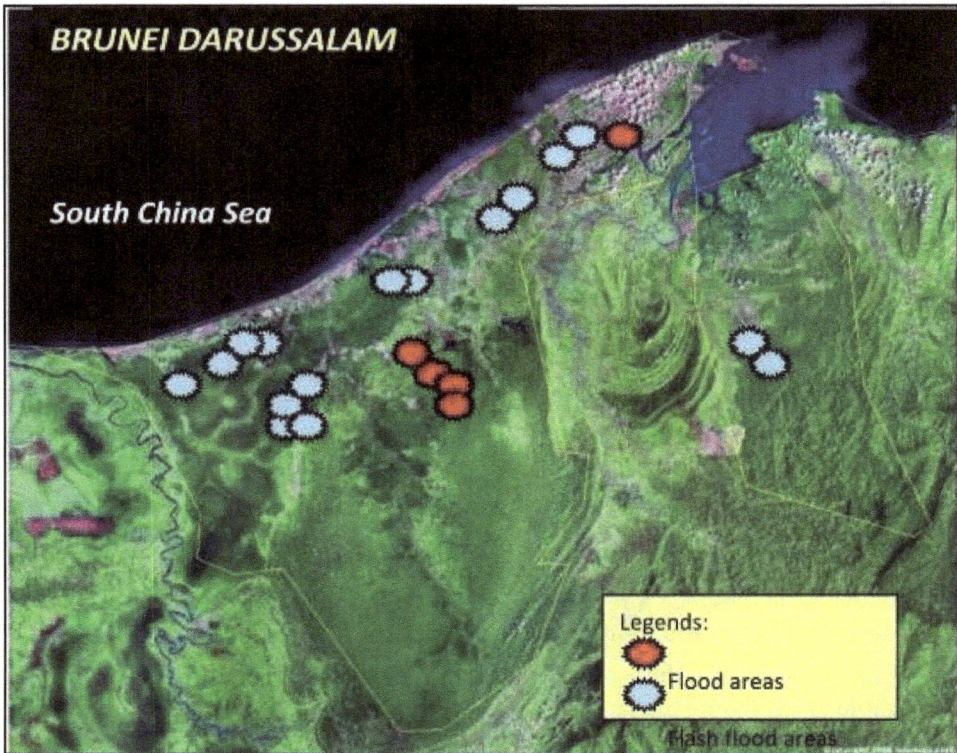

Figure 12.19: The River and Lowland Areas Affected by Flood and Flash Flood.

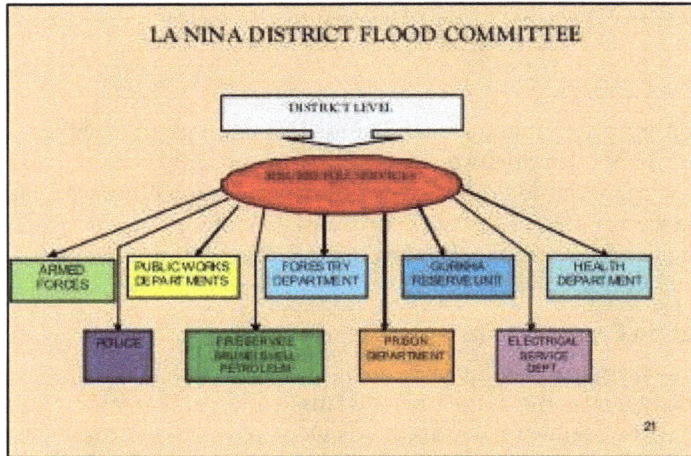

Figure 12.20: *La Nina* District Flood Committee.

The Affected Areas

During *La Nina* the heavy rain caused many parts of lowland areas overcome a flash flood throughout the country and the prone river banks over flooded for several days as the locations shown in Figure 12.21.

Technical Task Force

Unlike *El Nino*, *La Nina* only strikes on certain day or night and might only caused a short period of flash flood. During this event, the District Officer plays as the chairman for the task force and the front liners agencies consist of Department of Fire and Rescue, Armed Forces, Police other relevant agencies and the communities.

Landslide and Strong Winds

Besides the mainland areas, the other vulnerability sites which always confronted with the disaster are the Water Village. 'Kampong Ayer' or Water Village (Figure 12.21), also known as the Venice of the east, the man made disaster risk is mostly on fire destruction which will affect stilt houses built very close to each others (Figure 12.22).

Nevertheless, the other natural disasters which came from the meteorological effect are the landslide and the strong wind. Although such event was only on minor cases as being experience a few years back by the villagers, the government has taken serious concern and treats this as a priority for the purpose of public safety.

Some of the village houses located alongside the river bank (Figure 12.24) is vulnerable to overcome the land slide due to the hilly slope land especially during the down pour.

Strong winds and storms also affected some the roof trussed of the houses which mainly made up of wooden structures.

Figure 12.21: Aerial View of Water Village or 'Kampong Ayer'.

Figure 12.22: Aerial View of Water Village on Fire August 17, 2004.

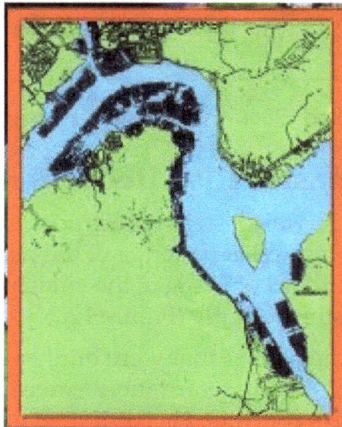

Figure 12.23: The Distribution of the Houses in the Whole of Water Village which Lies on.

High tide on the Brunei River is also one of the major problems for the water villagers.

Houses built near the hilly slope were vulnerable to the danger of the landslide and also affected the public accommodations. (Figures 12.24 and 12.25).

Figure 12.24: Affected House by the 1999 Land Slide.

Figure 12.25: Land Slide Affected Electric Cables and Passageway.

Disaster Risk Reduction

The Brunei Government has taken steps in restructure the new village settlement a few years ago which were located at the edge of the nearest water village for the purpose of promoting the standard of living of the village people who were affected by fires or other natural disasters recently (Figure 12.26).

With the new concept, the houses that were built, were designed by using the materials that are made up of the fire retardant, whereas the location of the houses were kept distance which were controlled according to the building regulation and laws (Figure 12.27).

Figure 12.26: New Settlement of Water Village.

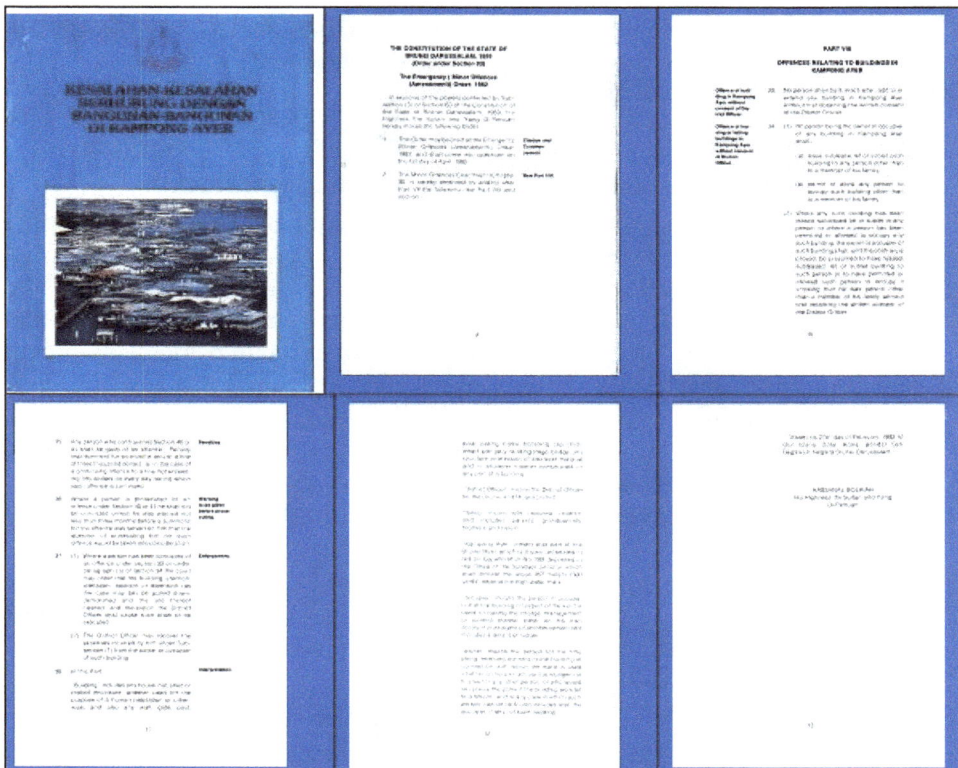

Figure 12.27: The Constitution of the State of Brunei Darussalam, 1959 (Order under Section 83), the Emergency (Minor Offences (Amendment)) Order, 1982.

In the mainland, the Government also have provided 23,000 hectares of houses for the people under the National Housing scheme throughout all the four districts at

Figure 12.28: The Houses Built by the Government Under the National Housing Scheme.

a reasonable cost, the types and design will be restricted under the government law and regulation imposed on the house owners. The government loan provided will be based according to their basic salaries (Figure 12.28).

Lesson Learnt

The following lessons were learnt from the past disaster incidents:

1. In global term, political, economic and social stability of this region depends significantly on bridging the gap between developing and developed nations. The world is facing a range of environmental crisis. The continuation and enhancement of international and regional cooperation and disaster assistance are key factors in the mitigation and containment of disaster effect.

2. In national term, the impact of such disaster had resulted in the direct loss of national assets in various forms. The diversion of national resources and effort was heavy in-order to achieve satisfactory recovery. A nation must develop a comprehensive approach to disaster managements having a strong National Disaster Management Policy.

3. In practical disaster management term, there is a need for an accurate and precise focus of requirements at various levels of disaster planning and that proper and appropriate training of staff is essential.

The Present Disaster Management System

Under the current system of disaster management, in house disaster and emergency plans are developed individually by respective government department having statutory responsibilities regarding disaster.

Disaster Management Orders 2006

Recognizing the shortcomings of the present system, Brunei Darussalam is embarking in putting effect of the Disaster Management Order 2006. In this order,

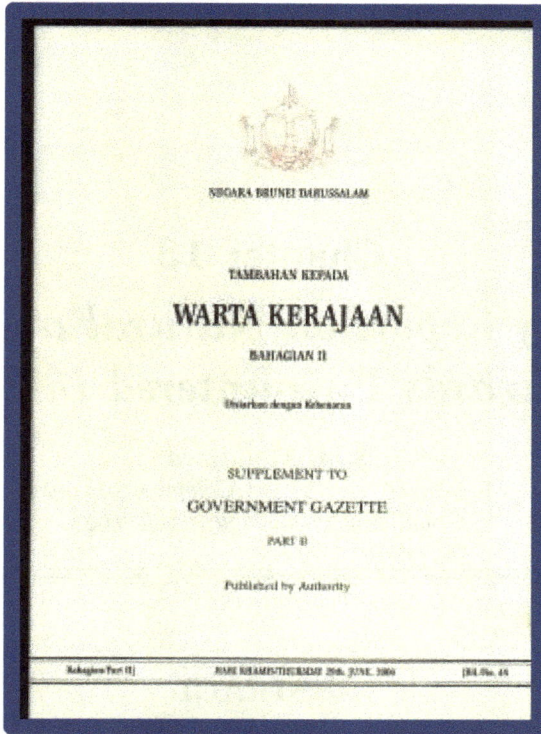

Figure 12.29: Disaster Management Orders 2006.

there is a clear definition of the national disaster management policy to deal with all aspects of disaster threat.

The main areas covered by this strong legislation are establishment of the National Disaster Council, National Disaster Management Centre, National Disaster Management Plan, Disaster Situation, Powers, Offences, Compensation and Regulations.

Conclusion

Climate changes have important implication for the economy of a country. The recent phenomenon of Asian Tsunami and hurricane season has shown that these events can result in devastating damage to both people and property.

The weather changes take place due to ocean and atmosphere interaction, and the two processes of *El Nina* and *La Nina* result in substantial changes in climate in various parts of the world. The changes in both temperature and rainfall would affect the people and thus their forecast helps to control and mitigate the effects.

Chapter 13

Country Report on Natural and Man-Made Hazards Encountered in Sri Lanka

S.M. Mohamed
Ministry of Resettlement, Colombo, Sri Lanka
E-mail: smarinamohamed@yahoo.com

ABSTRACT

After introducing geographical features of Sri Lanka, describes the disaster management efforts being implemented by the Sri Lankan government to mitigate the problems. Points out the natural disasters such as floods, droughts, landslides, cyclones, tsunamis and fires that Sri Lanka has normally faced, remain as threatening as ever. These disasters continue to cause grievous human casualties, economic and social loss and damage to the environment. Though we have learned to cope with these problems to some extent, we have neither eliminated nor contained them. Concludes that developing countries prone to natural disasters are not fully able to provide adequate resources for mitigation and prevention of disasters. Although international assistance is provided in most of the situations, these have been focused more on relief and less on mitigation or prevention The WCDR may review this matter and consider setting up an International Disaster Mitigation Fund, that would focus solely on prevention and mitigation aspects, and which can gainfully utilize the expertise and experience available with regional disaster management institutions and international NGO to assist the needy countries.

Introduction

Sri Lanka is an island located in the Indian Ocean at 5°55 and 9°51 North latitudes and between 79°41 East longitudes. It is bifurcated from the Indian subcontinent by a 32.5 km narrow strip of shallow water Palk Strait. It has compact land mass except for the Mannar Island in the northwest, the Jaffna peninsula in north and some satellite islands in the north of the peninsula. It has an area of 65,610 km of which

64,693 km is land 917km of large inland waters. 20,200 square kilometers (31 per cent) of land area is made for agricultural purposes and about 37,600km² (57 per cent) is underdeveloped land. The greatest length is 432 km from north to south and its width is 224 km east to west at its broadest part. The mountainous area is in the south central part of the country falling away on each side in successive steps to the sea. Five sixth of the area is at a level lower than 1000 feet and about three quarters below 500 feet. The Pidurutalagala is the highest peak at 2,524 m. The Sri Lankan climate is tropical with mean relative humidity varying from 60 per cent (in the driest areas) to about 95 per cent (at night, comparatively in the west areas). Relative humidity differs from 70 per cent during the day to 90 per cent at night. Mean yearly temperature ranges from 15°C in the mountains to 27°C in the lowlands. Temperature rarely exceeds 40°C. The highest temperatures are observed in April and May and the lowest in December through February.

Hydrologically, the country can be categorized into distinct wet and dry zones. The wet zone (South West) receives rain without pronounced dry periods but the dry zone (North East) only in the monsoon season. South West monsoons are generally experienced from May to September while North–East monsoons are experienced only from December to February. In addition, the whole island experiences inter monsoon rains in March to April and October to November period. Yearly rain fall varies from 1000 mm to over 5000 mm in South West of the island, less than 1250 mm in the North-West and South-East of the Island. The inter monsoon periods are the periods of transition from one monsoon to another and during these periods convective rain happens over many parts of the island. In addition, cyclones and depressions that form in the South Bay of Bengal also affect the climate over Sri Lanka especially during the months of October to December period.

Natural disaster events in Sri Lanka show a clear increase in the last two decades. Disasters experienced by Sri Lanka are categorized under floods, Cyclones, Drought, Landslides and Coastal Erosion. Sri Lanka has recorded earth quake events since 1614 with the most recent being in 1993 although their impacts have not been significant. Further more, health hazards, communicable diseases and environmental pollution events regularly happen in Sri Lanka. Over the past two decades the peoples of Sri Lanka have been suffering the rigors of social, political and economic upheavals due to political instability, insurgencies, militant uprising, and are caught in the grip of an armed conflict presently. The causative factors can be broadly identified as human made. Resources needed for growth and development have been diverted for containing these disasters. Uncertainty about the future has created feelings of insecurity and adverse impacts on the economy and the environment for development investment. This particular preventing situation in Sri Lanka has compelled the Government to take mitigation actions related to human caused disasters along with natural disasters.

Most of the major disaster events in Sri Lanka are due to bad weather conditions. Excessive rainfall, owing to upper atmospheric disturbances or low-pressure systems results in severe flooding and landslides. On the other hand severe drought conditions as a result of deficit rain fall are also not uncommon in the island with lower crop yield and power shortages. Sri Lanka is therefore seen to be exposed to risks from hazards like floods, landslides, cyclones, droughts, windstorms, coastal erosions and occasional seismic events. It is also prone to man made hazards like deforestation, indiscriminate coral, sand and gem mining, and industrial hazards, ethnic conflicts and occasional political violence. The severity and frequency of the disasters in Sri Lanka may not be on a high scale as one finds in some other countries in the region. However, the damages, hardships and the relocation arising from disasters, civil conflicts and political violence are indeed very grave for a small country like Sri Lanka. It is anticipated that the happenings of hazardous events may increase in the future due to the changes in demography, development patterns and the climate.

Mid year population of the island is 19.7 million according to the census of 2005 Average annual growth rate is 1.1 percent. Although the population density of the country is 314 persons per square kilometer, population density of Colombo is 3150 person per sq. km. Twenty three percent (23 per cent) of the total population are living in urban areas. Among the main cities (Colombo, Kandy, Galle, Jaffna and Kurunegala) of the island Colombo is reaching towards the million cities. The commercial centre is Colombo but the administrative and legislative capital is located at Kotte. Other major cities include Jaffna, Galle, and Kandy.

Since the colonial period, Sri Lanka economy depended on export agriculture till 1980s. At that time Tea, Rubber, and Coconut were the main export products. The situation changed after introducing a liberal economic policy to the country's economy in 1977. Access to foreign market and foreign investment was liberalized. Our country has been able to maintaining average economic growth rate of 5 per cent per annum during the period 1978-2005. Now more than 50 per cent of the economy is borne by the services sector. The highest contribution to the economy is services sector (56 per cent) and industry and agriculture contribute 26 per cent and 17 per cent respectively. GDP per capita of the country is US$ 1188 (2005).

Housing and sanitation and improvement of living conditions are important aspects of human settlement while achieving balanced development in keeping par with environment. In terms of household size, the average number of persons per housing unit in Sri Lanka has come down from 5.6 in 1971 to 5.2 in 1981 while the number of housing units had grown. More recently the average number of occupants per house has dropped further to 4.5 persons. About 75 per cent of the population now live in houses with more than three rooms compared to 31 per cent in 1953, and 46 per cent of the housing units had a floor area greater than 550 sq. ft in 1994 compared with 28 per cent in 1971. The housing stock in the country had increased from 1.5 million in 1953 to 3.8 million in 1994, with a rapid expansion since 1981. However in spite of concerted efforts to improve housing in the country, there is yet a scarcity of houses. It has been estimated that housing requirements would increase by about 330,000, between 2000 to 2005, of which 90,000 would be in the urban areas.

Disasters and Challenges for Developments in Our Country

Tsunami

Disasters are normally inter-linked with the people causing eminent damages to their properties and livelihood from the time the civilization dawned. Tsunami was the last spoken disaster in the Sri Lankan history. It devastated the entire coastal beds of Sri Lanka. The damages caused are still being rehabilitated in this Island.

Floods

Floods are the most common natural disaster in Sri Lanka and contribute 50 per cent of total disaster occurrences in the country, devastating people and property year after year. Between 1984 and 2006, the records say that floods have killed about 310 people and left almost 2.5 million homeless. There are 103 river basins in Sri Lanka of which 10 rivers such as Mahaweli, Kelani, Nilwala, Kalu and Gin are vulnerable to floods substantially. While heavy rainfall and run off the large volume of water from the catchments areas of rivers are the causes of much flooding, human activities such as deforestation, improper land use and absence of scientific soil conservation methods are contributing towards the increase in both frequency and impact of floods. Despite the districts of Colombo, Gampaha, Rathnapura, Kegalle, Galle and Matara are inundating to floods during the southwest monsoon, the districts of Ampara, Polonnaruwa, Trincomalee, Jaffna and Batticaloa are subject to floods during the northeast monsoon. Some relief and rehabilitation measures are in place during and after floods but no comprehensive legal and institutional arrangement to reduce the impact of flood hazards has been developed in Sri Lanka. The floods, which occurred in 1947, 1957, 1984, 1990, 1992 and 2003, were disastrous events the country ever faced.

Cyclone

The eastern, northern and north central regions are vulnerable to cyclone that occurs as a result of climatic changes of the Bay of Bengal. According to disaster history, cyclones are not frequent occurrences in the country but high gale and wind storms accompanied with the monsoon rains have frequently caused considerable damage to the people's livelihood and environment. Few of the cyclones in the past 70 years, have caused substantial damage to the society and economy. The disaster information shows that the 1978, cyclone was the most tragic disaster in Sri Lanka ever recorded. It claimed more than 1000 lives and 250,000 houses were damaged in eastern, northern and north central part of the country. The cyclone occurred in December 1999, has claimed 15 lives and damaged 77,000 houses in the districts of Trincomalee, Polonnaruwa, Anuradapura, Matale, Vavuniya and Mannar.

Droughts

Droughts are also a frequent occurrence in Sri Lanka and every year, for short durations droughts which are local significance are experienced by the people in some parts of the country. Despite regional significance droughts occur once in every 3 or 4 years, the people and economy have been hit by severe drought of national significance once in 10 – 15 years consequently. Least rainfall, deforestations, improper land use and unplanned cultivation patterns are the main causes for the drought in some parts of the country, such as north-central, south-eastern, north-western and eastern regions. Major droughts occurred during the period of 1953-1956, 1974-1977, 1981-1983 and 1995-1996, causing substantial damage to the economy and plaguing development efforts of Sri Lanka. The period of 2000-2002, 1.6 million people living in Southern and North Western dry zone areas were experiencing severe drought.

Landslides

Landslides, which are often triggered by intense heavy rains, have caused imminent damage to people and development activities of the country especially during the last two decades. Changes in land use and cropping patterns, increased demand for land that pushes settlements into unstable areas, alterations in normal drainage flows, improper slope cutting techniques have caused increasing of the landslides in hilly areas substantially. It is identified that seven districts in central and southwestern region covering over 12,000 square kilometers land area of the country are highly vulnerable to landslides. The landslides occurred in 1989, 1993 and 2006 ever recorded hazardous events in the Disaster history of Sri Lanka have triggered severe damage to people and property.

Sea Erosion

Sea Erosion that can be seen as a disaster during the past two decades in the country have caused extensive damage to the western coastal line and to people's livelihood in Sri Lanka. It is estimated that over 50 per cent of the shoreline of Sri Lanka is subject to or threatened by sea erosion. The most vulnerable areas are those between Kalpitiya in the Northwest and Tangalle in the south. Off shore mining of corals, removal of sand from the beach, clearing of mangrove swamps and destruction of coastal vegetation have resulted in sea erosion in the country.

Disaster Management

The concept of the disaster management is very much an ongoing national requirement which is important to government and officials who are involved in disaster management activities. It has a special significance today as for the increasing dangers to the world environment due to climate change, sea level rise and ozone layer depletion etc. in the respective countries.

The significance of disasters some times comes under question? Why do we need to be concerned so much on the lesson from natural disasters? Disasters have been with us as long as the world history of mankind is recorded. Generations of people have had to withstand disasters. They have suffered the consequences and recovered from them and the lives have continued on. But in today's' environment, we need to pay more attention on disasters as the disasters have become hazardous not because of natural disasters but of manmade disasters and then man makes it worse. Therefore, certain factors need to be considered in relation to the challenges which face disaster management.

Natural disasters such as floods, droughts, landslides, cyclones, tsunamis and fires that Sri Lanka has normally faced, remain as threatening as ever. These disasters continue to cause grievous human casualties, economic and social loss and damage to the environment. Though we have learned to cope with these problems to some extent, we have neither eliminated nor contained them. We find it difficult to make ends meet.

Increased social violence has drastically affected Sri Lanka and the community as well. The terrorism, civil unrest and conflict with conventional arms have become hazardous. This has inflicted intolerable burden on the government and the society whose existence is already precarious because of poor economic and social condition in Sri Lanka. This, in turn has produced additional strains on international assistance sources, thus diluting global counter – disaster effort and capability.

One of the main factors which should be considered is the relationship between the contemporary disaster threat and the losses it may impose. The fact that the more country develops and the more assets it builds up, the more it stands to lose. It follows that any action which can be taken to reduce disaster – related loss must be seen as logical and desirable in cost – benefit terms. This applies to all countries to try to develop and maintain an effective disaster management capability appropriate to their needs. It underlines the necessity for coordinated international action in order to strengthen all aspects of disaster management, where ever it is possible.

In global terms, unless disaster can be mitigated and managed to the optimum extent possible, it will continue to have a dominating effect on the future. Disaster mitigation should be regarded as an important tool in successfully coping with these crises. The political, economic and social stability of the world depends significantly on bridging the gap between developing and developed nations. The mitigation and containment of disaster effects on the developing nations now and in the future, is an important asset towards bridging this gap. The continuation and enhancement of international disaster assistance is also a key factor.

Correctly applied, such assistance can help to provide a desirable bonding between nations and thus produce welcome and beneficial long-term results. The impact of disaster usually results in two major setbacks

1. The direct loss of existing national assets.
2. The diversion of national resources and efforts, away from ongoing subsistence and developments, in order to achieve satisfactory recovering.

This indicates that the nation needs to develop a comprehensive approach to disaster management. To be effective this comprehensive approach clearly needs to cover all aspects of the disaster management cycle and needs to include an appropriate balance of prevention, mitigation, preparedness, response, recovery and disaster related development. It is true that improvements can result from disaster. How ever this does not reduce the need for a comprehensive approach. In fact, such an approach because of it's inter relationship with national development, is more likely to ensure potential benefits.

To be effective, therefore disaster management needs to be implemented as a comprehensive and continuous activity not as a periodic reaction to individual circumstances. Consequently national representatives like us charged with disaster management responsibilities have to deal with a wide range of policy planning, organizing, operational implementation, monitoring and evaluating matters.

Sri Lankan Government is at present in the process of finalizing the legal provisions for Disaster Management through the Disaster Management Counter

Measures Bill that has been referred to Parliament for sanction. National Disaster Management Council will be apex legal body for the formulation of work plans and programme according to the Bill. It consists of policies and an improved institutional arrangement from national to village level. This Bill would be active as a live document once it is enacted by the parliament. Floods, Landslides, cyclones, tidal waves and droughts are the normal natural disasters that our country has been facing. The country's civil war, accidents, industrial hazards and environmental degradation are to be reckoned as the human made disasters. Not only the losses of lives and livelihood, but also heavy economic losses did take place. Within the period of one decade or so Sri Lanka experienced many major disasters wherein many were killed and 27.5 million people were affected in cases other than the civil war. The flood and landslides occurred in 2003 claimed 252 lives and about 26,000 houses were estimated damaged. The man-made disasters have killed about 64,000 people and 20 million people affected since the July 1983 communal riots.

National Policies of Sri Lanka

National Policies of Sri Lanka that are to be addressed in this connection are as follows in a concise form:

☆ Government will protect the life of the community state/private property and the environment from disaster as a core responsibility of government to the community at large;

☆ Partner non-governmental organizations, private sector, media and community group in all aspects of disaster management and mitigation;

☆ Ensure that disaster mitigation activities promote and contribute to national development and that development projects contribute to disaster mitigation;

☆ Ensure urban planning;

☆ Organize and co-ordinate the effective use of resources for preparedness, Prevention, response and relief and reconstruction and rehabilitation;

☆ Maintenance and development of disaster affected areas;

☆ Develop and implement programme for public awareness and training to help People protect themselves from disasters. Provincial Councils, other levels of Local Government, agencies and departments of the National Government and other organizations will develop plans within the overall framework of the National Plan;

☆ Encourage and support the introduction and application of improved professional practices in the areas of agriculture, land use planning, watershed management, coastal areas, construction and maintenance to reduce vulnerability and losses;

☆ Foster scientific, technological and engineering endeavours (*e.g.* landslide hazard Mapping) as tools for sustainable development;

☆ Promote the shifting of emphasis to pre-disaster planning, preparedness and mitigation while sustaining and further improving post-disaster relief, recovery and rehabilitation capabilities;

☆ Support the enhancement of local capability and capacity to manage risk and apply disaster management and mitigation practices;

☆ Encourage participation of NGOs mass media, private institutions and individuals and will assist in directing private donations to proper recipients in affected areas;

☆ Ensure integration of disaster prevention and preparedness in the national as well as sub -national planning process.

Disaster Mitigation

The following principles are widely recognized as providing a valuable guide to disaster mitigation which is being administered in our country.

Initiation

☆ Disasters offer unique opportunities to introduce mitigation measures.

☆ Mitigation can be introduced within the three diverse contexts of reconstruction, new investment and the existing environment. Each presents different opportunities to introduce safety measures.

Management

Mitigation measures are complex and interdependent, and they involve widespread responsibility. Therefore, effective leadership and co-ordination are essential to provide a focal point.

☆ Mitigation will be most effective if safety measures are spread through a wide diversity of integrated activities.

☆ "Active" mitigation measures that rely on incentives are more effective than passive measures based on restrictive laws and controls.

☆ Mitigation must not be isolated from related elements of disaster planning such as preparedness, relief and reconstruction.

Prioritization

Where resources are limited, priority should be given to the protection of key social group, critical services and vital economic sector.

Monitoring and Evaluation

Mitigation measures need to be continually monitored and evaluated so as to respond to changing patterns of hazards, vulnerability and resources.

Institutionalization

☆ Mitigation measures should be sustainable so as to resist public apathy during the long periods between major disasters.

☆ Political commitment is vital to the initiation and maintenance of mitigation.

Application of Mitigation measures

☆ Strengthening building to render them more resistant against cyclones, floods or landslides, drought.

☆ The incorporation of hazard resistance in structures or procedures to be followed in new development projects.

☆ Planting certain kinds or varieties of crops that are less affected by specific kinds of disaster.

☆ Changing crop cycles so that crops mature and are harvested before the onset of the flood or cyclone season.

☆ The adoption of land-use planning and controls to restrict activities in high risk areas.

☆ Economic diversification to allow losses in one sector to be offset by increased output in other sectors.

Certain Other Aspects in this Connection are as Follows

☆ There may be longstanding acceptance of disaster risks by governments and communities, who may feel that traditional measures (taken over many years) are adequate. Such measures may include the positioning of population sites and traditional building practices. There may therefore be some built in reluctance to accept new methods of mitigation.

☆ Some mitigation measures may be costly for example; enforcement of buildings codes is likely to increase the cost of buildings. This in turn may reflect, in various ways, on costs and prices, and may therefore be opposed within the community.

☆ Higher priorities given to other major national programs (health, education etc) may make it financially difficult to implement mitigation programs

☆ Political considerations may rule out or restrict mitigation programs. If such programs are extensive and (through land use restrictions and enforcement of building codes, etc) unduly interfere with living conditions and standards, governments may become unpopular. Thus, governments may not enforce mitigation programmes to full effectiveness, in the interests of retaining political power.

☆ Aspects modern progress and development may affect mitigation programs. For instance, international standards in various fields may dictate that governments must undertake mitigation measures (mainly perhaps in the form of safety measures). In such cases, governments may have little or no choice. This may therefore mean that other desirable mitigation programmes will have to take lower priority.

☆ Lack of or insufficient appropriate mitigation measures may have an adverse effect on the ability to cope with disaster situation. For instance, inadequate mitigation measures may cause a significant overload on response operations and result in the latter being only partially effective.

☆ If counter disaster planning is inadequate, the effectiveness of mitigation may be seriously reduced. For example, it may be possible to mitigate the effects of a disaster situation by undertaking a precautionary evacuation of people before the disaster strikes. However, if evacuation plans and other counter-disaster arrangements are not already in place, such an evacuation may not be possible. Even worse, if such an evacuation is attempted without adequate plans, the risks to the people concerned may even be increased. Similarly, if planning is inadequate, large numbers of people may, as a mitigation measure, be housed in unsafe communal buildings (schools, churches, etc). In the past, in some countries, this has led to considerable loss of life, due to buildings collapsing during disaster impact.

☆ In post-disaster analysis and review, insufficient attention may be given to mitigation measures. This can have severe repercussions in future disasters.

☆ Inadequate standard of community self-reliance and self-help may adversely affect successful mitigation because even elementary precaution (such as ensuring an emergency food supply or being prepared for evacuation) will not be taken.

Requirements for Effective Mitigation

Preventive actions for effective mitigation tend to be the responsibility of senior levels of government and of senior management in the private sector. To some extent this applies to mitigation, where measures such as building codes and land use regulations usually emanate from major policy decisions. However, for mitigation, it is worth noting that lower levels of government may play a greater part. For instance, precautionary evacuation as a mitigation measure tends to be the responsibility of local government or the local counter disaster committee. Also precautionary evacuation may even result from a local community decision or reaction.

Requirements for effective mitigation may include some or all of the following.

☆ A clear and comprehensive national disaster policy with addresses all aspects of disaster management and ensures that mitigations are given proper consideration and priority.

☆ Adequate assessment and monitoring of disaster hazards and vulnerabilities, so that the need for mitigation measures is accurately identified and defined. Indeed, effective vulnerability analysis is a primary prerequisite for mitigation programmes and annexure A deals with this subject in detail.

☆ Adequate and accurate analysis of all reasonable mitigation projects. In this regard, it is especially important to achieve sensible gain/loss comparisons for instance, whether by instituting mitigation programs the nation and community are going to gain more (bearing in mind the costs and restrictions involved) as against the losses which might arise if nothing is done.

☆ Readiness on the part of governments to institute and carry through appropriate mitigation programmes.

☆ Appropriate consideration of mitigation measures in National Development Plans, including the immediate and long term cost benefit implications of taking or not taking mitigation action.

☆ A basis of organization and planning centred on a permanent disaster management centre or section. The existence of such a section is vitally important in the overall disaster management sense because on behalf of government, the centre/section should keep a constant watch on disaster management. Thus in coordination with other agencies it is able to identify the need for various mitigation measures as they may arise. It is then the responsibility of the centre/section to advice government on needs for mitigation programs, and the priorities, which should apply.

☆ Insistence by the disaster management centre/section (on behalf of government) that an effective post-disaster review is undertaken after all major disaster events. This review must include advice to government on whether, as a result of a particular disaster, mitigation measures are adequate or whether additional measures are needed.

☆ Recognition that mitigation measures may originate from all levels of government not only from national level. This is important because for instance, the "disaster front" is usually at local government (or community) level. Thus, from this level the need for mitigation measures may be more obvious than from higher levels.

☆ Specialist programmes which may assist in the development of large-scale mitigation measures. For instance, agricultural programmes, which assist farmers and others in mitigation of, crop losses.

☆ Adequate public awareness and education programs, in order to assist communities in playing their appropriate part in mitigation measures.

☆ Support for traditional measures of mitigation, where these may be of use in the overall disaster management sense.

☆ Support also for the development of self-reliance and self help at community level, because these aspects can often provide useful support for mitigation concepts.

Non structural mitigations like legal frame work, incentives, training and education can also be mentioned here. The four target groups as indicated below are especially important:

☆ Public officials who play a vital role in disaster management. Appropriate training modules should be incorporated in their career path training programs and opportunities provided to them to attend specialist courses.

☆ Technical students whose professional education may also include disaster mitigation courses.

☆ Small builders and craftsmen who may be given on the job training in simple mitigation practices.

☆ School children should be introduced to simple mitigation measures in the context of environmental studies, natural science or geography classes. (Fiji and Philippines have developed appropriate curricula and materials which may be of assistance to other countries).

Public Involvement

In addition to general awareness, certain particular areas of public involvement are necessary for effective implementation of mitigation programs. These include:

☆ A good public knowledge and understanding of local hazards and vulnerabilities.

☆ Public awareness of the kind of mitigation measures which can be applied

☆ Public participation in community preparedness programmes.

Governments can substantially assist public awareness of safe mitigation practice by ensuring that their own public buildings (such as post offices, schools, hospitals, government offices) and services are built to high safety standards. This will also help to ensure that designers, builders and engineers gain experience in safe construction and at the same time, contribute to a safer environment.

The strengthening of a country's or communities' social structure can enhance disaster mitigation capacity. Such strengthening is however, difficult to achieve. Three possible ways are to extend normal development as follows:

☆ First through institution building. Organizations that serve as coping mechanisms can be identified and strengthened. A deliberate effort can be made to increase their institutional capacities and skills, thus enhancing their ability to deal with a crisis.

☆ Second, through increasing the number of coping mechanisms within a country to community. Developing formal intuitions and linking them to outside resources establish means established for intervention and the provision of assistance.

☆ Third, through encouraging actions that promotes co-operation among different groups within society. Such co-operation can considerably reduce the social impact of disaster.

In their development activities, both government and non-government agencies should be careful to avoid actions that will further increase or institutionalize a society's vulnerability. It is especially important to identify institutional dependency relationships, particularly those that may be increased in a disaster situation, and work to eliminate them. By increasing self-sufficiency, agencies may improve the ability of families and communities to cope with disaster. This can be a mitigating factor and help to speed recovery. Strong institutions can play a vital role in various aspects of mitigation, such as promoting public awareness programmes, training at community levels and monitoring hazards and vulnerabilities.

Warning Systems

Various modern developments have significantly improved the ability of disaster management authorities to provide effective warning of impending disaster. Better warning systems have, for instance, been instrumental in evacuating vulnerable groups, moving livestock to safety and mobilizing emergency services and resources. These and associated matters are covered in detail in Annexure B, however in the particular context of mitigation, three are underlined here.

☆ The steps between the issuing of warning and the taking of action relevant authorities or vulnerable people are critical.

☆ Evacuation should only be ordered when there is virtual certainty of hazard impact, a false evacuation order for a hazard that does into materialize can destroy public confidence in the warning system and neutralize several years of preparedness planning.

☆ To the extent possible, the dissemination of warning should use duplicate systems to ensure effectiveness. For example, radio message backed up by siren warnings, warning flags backed up by house to house visits by local wardens.

Mitigation in Agriculture

Various measures can be applied in agriculture to mitigate the effects of disaster. These include.

☆ The planting of shelter breaks, comprised of trees and shrubs, to reduce wind effects.

☆ Crop diversification.

☆ Adjustments to crop planting/harvesting cycles.

☆ Food storage programs to insure against shortage arising from disaster.

Structural Measures

Non-structural mitigation measures may need to be complemented by **structural measures**. In the case of flood-prone areas, embankments, regulators, drains or by pass channels can be provided where appropriate, to protect areas from damage by floods. Techniques to mitigate the effects of landslides and floods on structures also exist. Structural mitigation measures may apply to both engineered structures and non-engineered structures.

Engineered structures involve architects and engineers during the planning designing and construction phases. They may include buildings ranging in scale from simple dwellings to multi storey office blocks as well as infrastructure such as electricity pylons to dams, embankments, ports, roads, railways and bridges. While professionals are trained to plan, design and supervise the construction of buildings and infrastructure to achieve necessary structural safety standards, they may need additional training to incorporate mitigation practices into their design of structures resistant to seismic shock, storm winds or floods. The application of sound technical principles is achieved through.

☆ Site planning,

☆ Assessment of forces created by the natural phenomena (landslides, Tsunami and flood),

☆ The planning and analysis of structural measures to resist such forces,

☆ The design and proper detailing of structural components,

☆ Construction with suitable materials, and

☆ Good workmanship under adequate supervision.

Most countries have building codes for engineered construction. These codes provide general guidelines for the assessment of forces and further analysis, appropriate design methodologies and construction techniques. If a country does not have a building code which specifies design and construction requirements for earthquake and wind forces, such a code should be formulated as soon as possible, technical personnel trained in its use and enforcement ensured. The quality of construction is as important as good analysis and design. Good workmanship must be encouraged by appropriate training and supervision to achieve better performances.

Non-engineered structures are those constructed by the owners themselves or by local carpenters and masons who generally lack formal training. Such structures mainly comprise simple dwellings and public buildings, built with local materials in the traditional manner. In some disasters, high casualties and economic losses can be attributed to the failure of non-engineered structures. The improved designs vary according to the many different traditional ways of building that suit various cultures, climates, available skills and building materials.

Another important aspect of increasing the safety of non-engineered structures is to try to ensure that they are not built on hazardous sites such as steep slopes subject to landslides, floodplains subject to flash floods or riverbank erosion, or coastal areas exposed to storm surges. However, people often do not want to leave their traditional homes and the area in which they have been living for generations, even though the location may be hazard prone. Economic pressures may also induce people to settle in hazardous areas. Wherever practical, incentives should be offered to attract people out of hazardous locations. Alternatively, consideration may be given to substituting appropriately engineered structures, where this may be practical and economic, or introduce mitigation measures in non-engineered construction so as to enhance their safety.

Conclusions

Information exchange and early warning among all countries can have significant preparedness and impact reduction effects by facilitating the establishment of bilateral and regional agreements/MOU among countries in this regard.

More decentralization of the existing institutions such as the ADPC (Thailand), CDMM (India) and ADRC (Japan) by setting up national centres in the regional countries can improve this situation.

It would be of great benefit particularly to the developing countries prone to natural disasters, if technical assistance were available specifically to examine and help improve the overall disaster mitigation structure from the national to community levels and related aspects in the country to mitigate vulnerability, so that the country is better prepared for natural disasters. Often developing countries prone to natural disasters are not fully able to provide adequate resources for mitigation and prevention of disasters. Although international assistance is provided in most of the situations, these have been focused more on relief and less on mitigation or prevention The WCDR may review this matter and consider setting up an International Disaster Mitigation Fund, that would focus solely on prevention and mitigation aspects, and which can gainfully utilize the expertise and experience available with regional disaster management institutions and international NGO to assist the needy countries.

Notes

About 50,600 families affected by Tsunami in Sri Lanka are yet to receive their permanent houses - RADA Reports.

According to the information released Relating to Sri Lanka by the Disaster Management Centre.

Deaths	30957
Missing Persons	5644
Injured	15196
Affected Families	202742
Displaced Families	84735
Internally Displaced Persons	396170
Fully Damaged Houses	78047
Partially Damaged Houses	41097

Chapter 14

The Development of National Early Warning Systems at BMG Indonesia

Sri Woro B. Harijono
National Meteorological and Geophysical Agency (BMG),
Jakarta Pusat–10720, Indonesia
E-mail: sriworo@bmg.go.id

ABSTRACT

The paper gives an overview on the various components of the Primary Warning System being used in Indonesia including infrastructure, hazard monitoring, and related activities like international collaboration, weather services, watches and advisory and dissemination of the information. It describes in detail the various activities of BMG in this direction. It also points out that development of integrated meteorological observing network to acquire surface and upper air data, as well as utilization of reanalysis/ prediction data, has raised the performance of BMG on providing weather, climate services and earthquake information and tsunami warnings. In the future, BMG will develop state-of the art Numerical Weather Prediction (NWP) model and climate forecast technologies to forecast the changes of weather and climate.

Introduction

Meteorological-geophysical induced disasters have been recurrently occurred in Indonesia. Devastating tsunami in Aceh and Pangandaran are the examples of geophysical – induced disasters, whilst whirl wind, strong wind, flash flood, and heavy rainfall are all meteorological – induced disasters. They caused hundred thousand deaths, destroyed million houses and resulted in direct economic loss. In response to those heavy impacts, establishment of multi hazard early warning system has been developed by the Indonesian Government since October 2005.

By law and in accordance with its tasks and functions, the scope of responsibility of BMG is mainly in the area of providing weather – climate – earthquake – tsunami information to address the need of the communities at large. Accordingly BMG formulated and developed two main systems, **Meteorological – Climatological Early Warning System (MC EWS) and Indonesia Tsunami Early Warning System (InaTWS)** as shown in Figure 14.1.

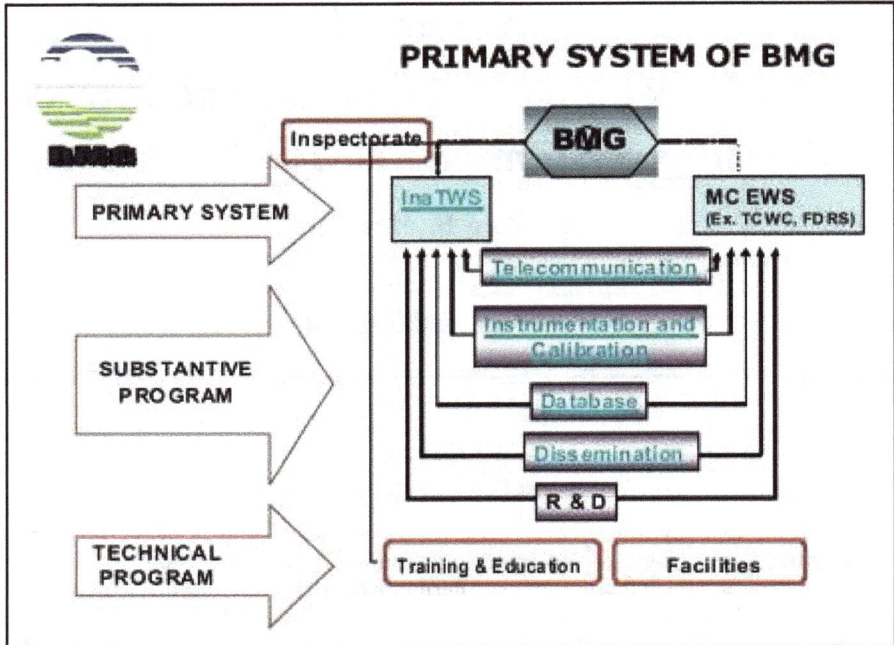

Figure 14.1: Primary Programme of BMG.

In Indonesia, National Coordinating Agency for Disaster Mitigation (BAKORNAS PB) is responsible for laying down policies, plans and guidelines as well as coordinating and activating all concerned-institutions that make contribution in accordance to the respective task and function.

Various warnings from BMG are transferred directly to BAKORNAS through private radio link for actions. Receiver system consists of server, modem and software application has been installed in BAKORNAS operation room by BMG, in order for BAKORNAS to be able to disseminate warnings and motivate/ activate institutions responsible for emergency measures.

In the provincial level, the BAKORNAS represented by Local Emergency Management Unit–SATKORLAK and further in the district level, coordination of disaster mitigation activities is tackled by SATLAK.

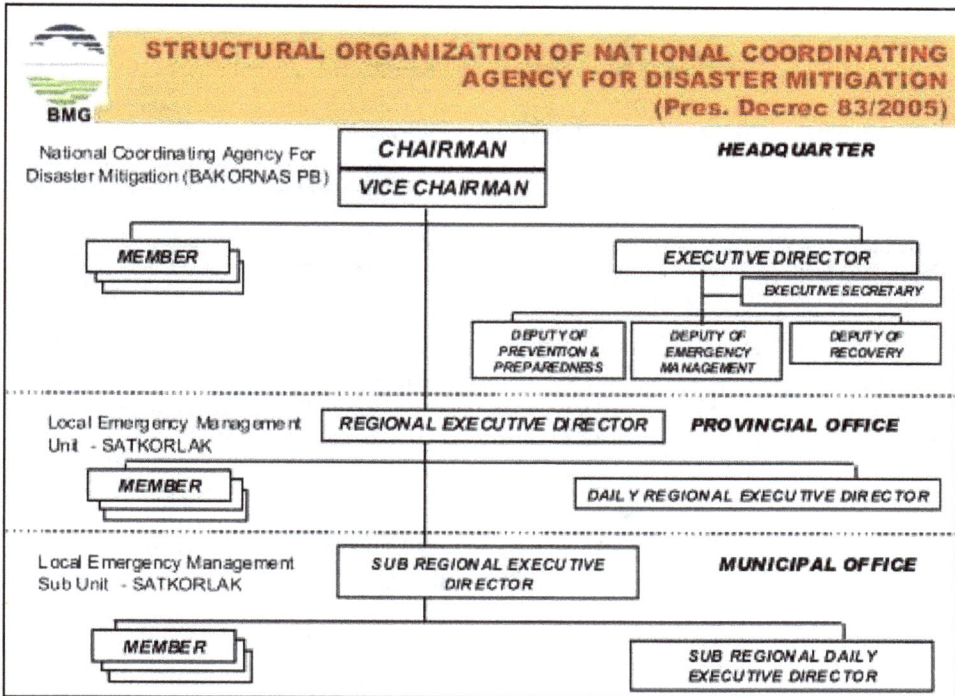

Figure 14.2: Structural Organization of National Coordinating Agency for Disaster Mitigation.

Primary Systems of BMG

Meteorological–Climatological Early Warning System (MC EWS)

The answer to addressing the need of community at large in hydrometeorological disaster advisory and warnings may lie in the development of Meteorological – Climatological Early Warning System (MC EWS).

As shown in more detail in Figure 14.3, MC EWS refers to the activities of **monitoring, processing, analyzing and producing and disseminating meteorological – climatological information.**

All of the data, (observational and re-analyzed forecast data) are collected in the **Transmet System** and **sent** to the integrator server. From integrator server, the specific data *e.g.* maritime-meteorological data are further transferred to the **"client"** system for various processing to produce meteorology maritime prediction such as wave height and sea current prediction. The weather forecasts are disseminated to the users by various medias: internet (email, web), telephone line (phone and fax), and mobile phone (Short Message Services / SMS).

Indonesia Tsunami Warning System (InaTWS)

Specifically, with regard to TWS, there is a general concept of End to End Disaster Risk Management (Figure 14.4). This concept has been frequently promoted and

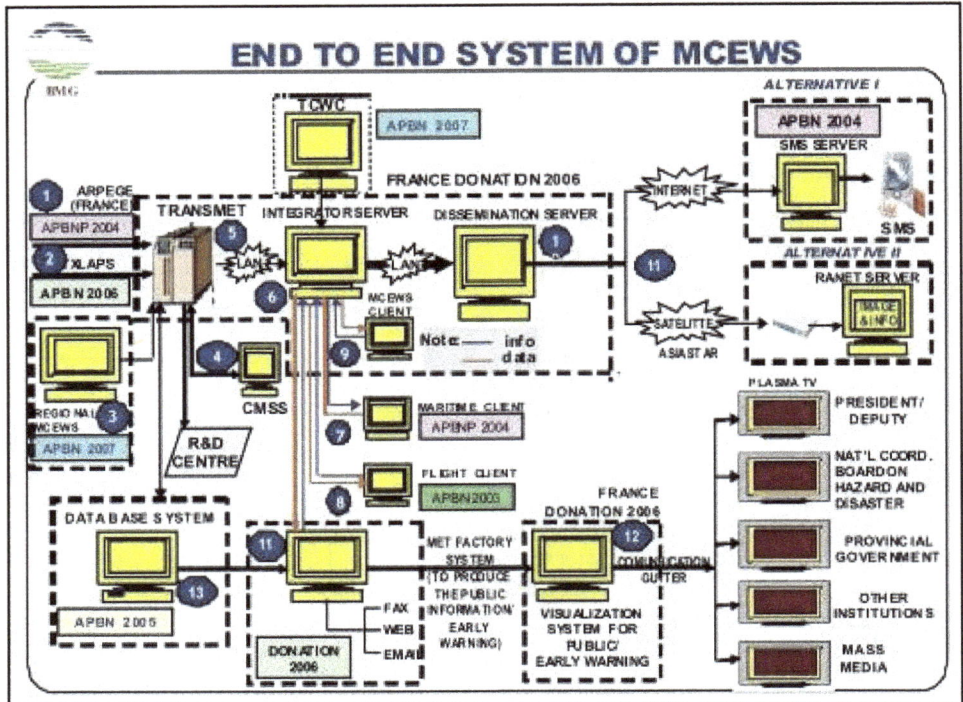

Figure 14.3: End to End System of MC EWS.

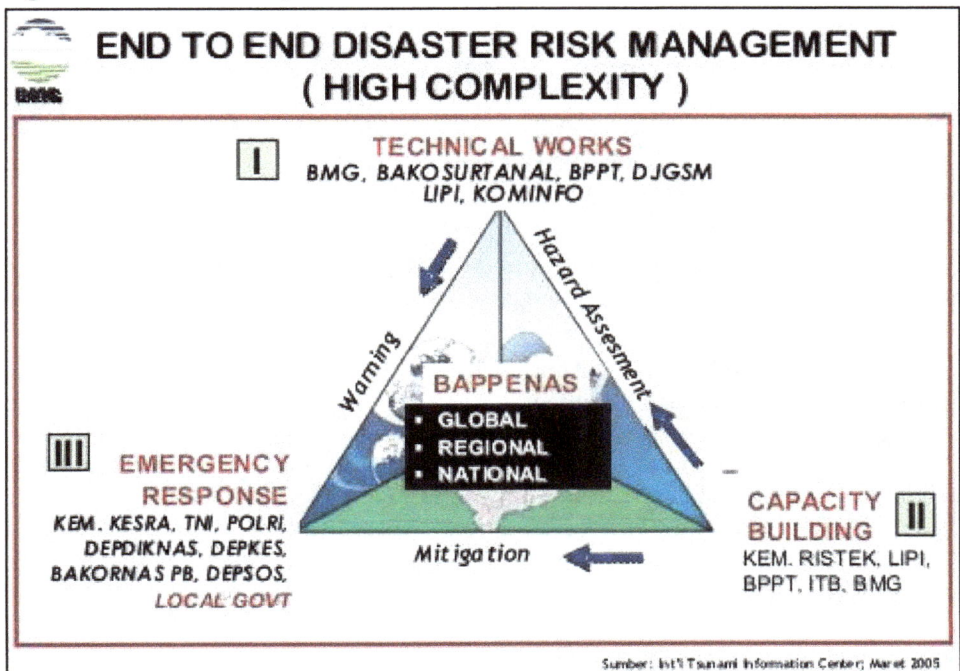

Figure 14.4: End to end Disaster Management.

highlighted in international meetings, which has been referred to be implemented in TWS development. There are three project components which covers the ability: **Technical works; Scientific capacity building (modelling – simulation – training – workshop); Emergency Response, Preparedness and Rehabilitation**.

As laid out in Figure 14.5, BMG has been appointed as the leading institution in the project component of technical work. This work is ranging from site survey – deployment of the sensors – data collection and process–producing information, as well as dissemination of information to the emergency institutions such as National Police Headquarter, National Radio Station and TV Stations. All of these technical works are for the purpose of civil evacuation as required by the municipal and district government.

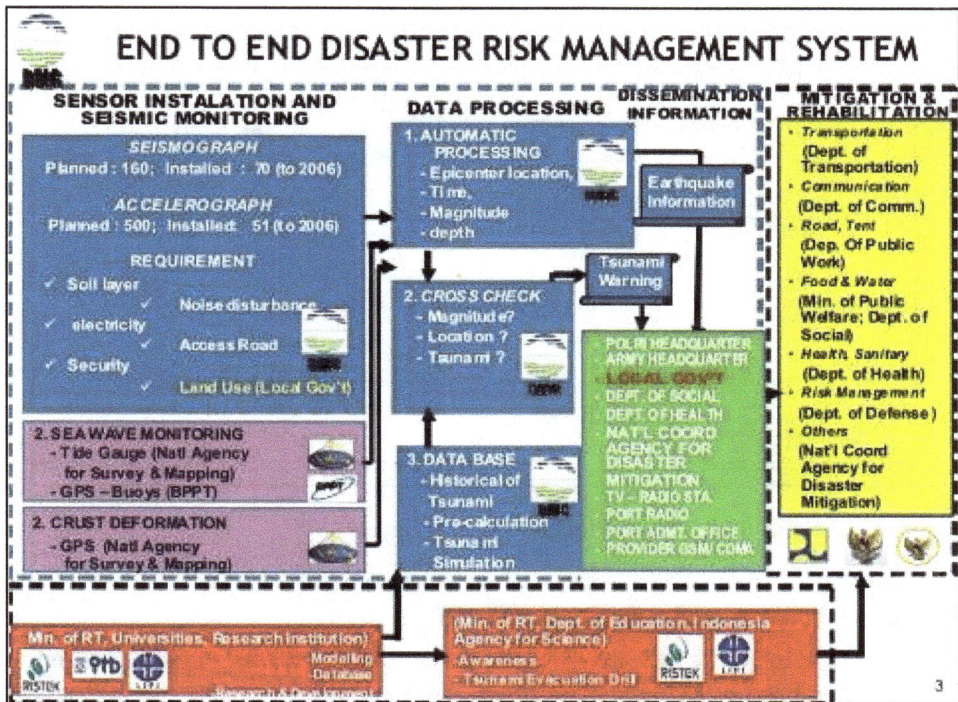

Figure 14.5: End to End Disaster Risk Management System.

Infrastructure of Early Warning System

The two main systems: Meteorological – Climatological Early Warning System (MC EWS) and Indonesia Tsunami Warning System (InaTWS), consist of sub-system including **monitoring system, processing system, as well as warning dissemination and a communication system**.

Meteorological–Climatological Early Warning System (MC EWS)

An integrated meteorological observing network is being developed particularly for MC EWS, including Weather Doppler Radar – Automatic Weather Station – Rain Gauges and Satellite Ground Receiver.

The total number of weather stations is 120 locations, in which 59 station are functioned as basic stations which has the obligation to exchange its data internationally. The network of existing weather radar, Automatic weather stations, and ground satellite receiver is as illustrated in Figures 14.6–14.8.

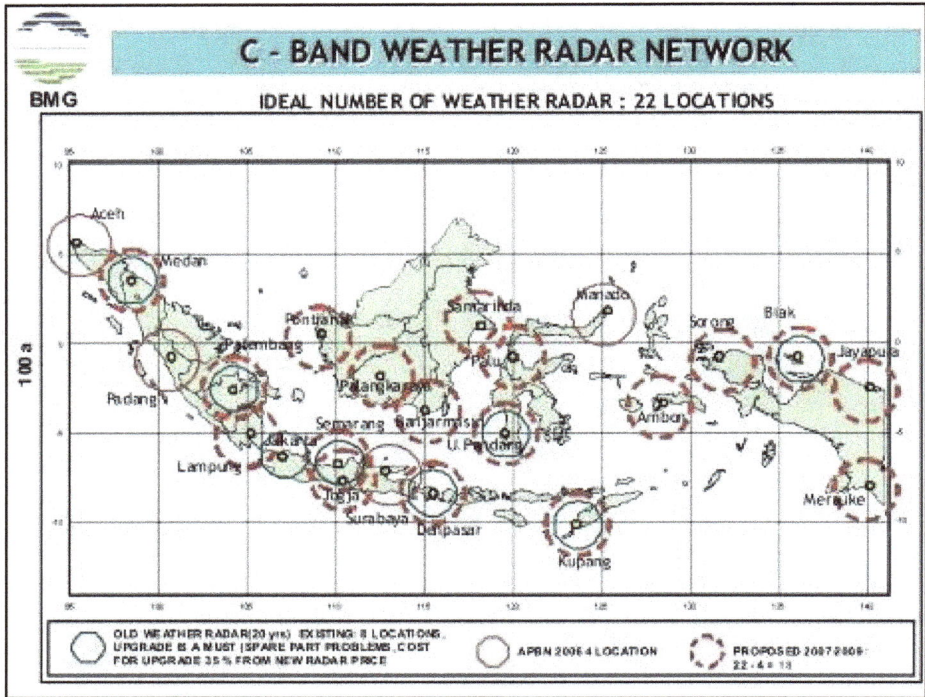

Figure 14.6: Network of C-Band Weather Doppler Radar.

Tsunami Warning System (InaTWS)

As shown in Figure 14.9, there will be 10 Regional Centre and 1 National Centre in Indonesia. Each of Regional Centre will be equipped by sensors, an automatic processing system, dissemination system as well as development on scientific capacity to increase the accuracy of analyses–processing.

The number of facilities that has to be completed by end of 2008 is 180 seismographs and 500 accelerographs.

Hazard Monitoring System

Meteorological–Meteorological Early Warning System (MC EWS)

An integrated meteorological – geophysical observing network has been set up in BMG Headquarter. The Meteorological–observing network includes Rain Gauges–Doppler Radar, Automatic Weather Stations, and Satellite Ground Receiver.

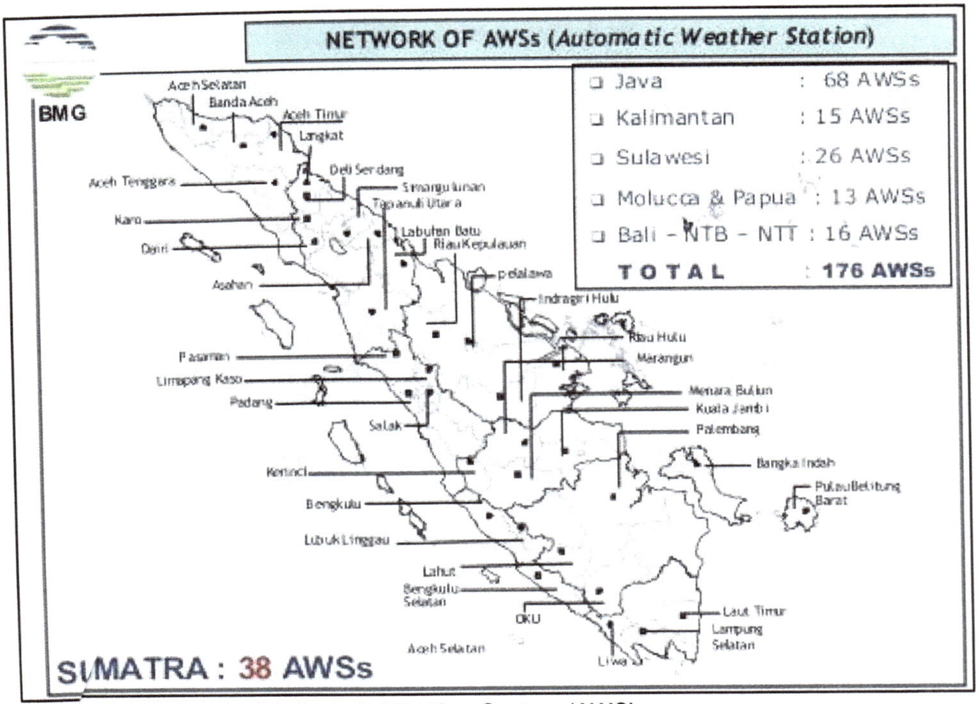

Figure 14.7: Network of Automatic Weather System (AWS).

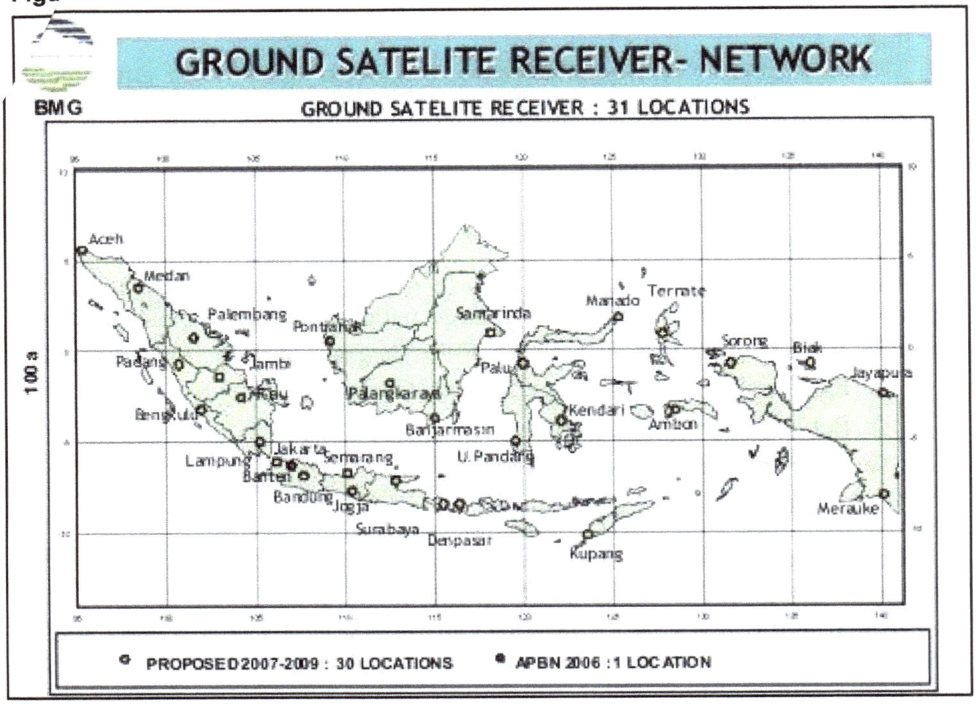

Figure 14.8: Network of Ground Satellite.

FDRS Indonesian Initiative, was completed in March 2003. The second component, the FDRS Innovation Initiative, is intended to overlap with the Indonesia Initiative and continue until June 2005. Key achievements in 2000–2003 are summarized below by project outcome.

A sequential approach is being used to implement two local pilots, beginning with a province in Sumatra. During the first local pilot, the CFS was leading the adaptation, operator training and output-based application activities. During the second local pilot, these activities were directed by Indonesian agencies with support from CFS. Implementation of the FDRS initiative in Indonesia will be through four inter-related programmes: Adaptation Technical assistance to adapt and adjust components of existing FDR systems for local tropical conditions to support fire prediction, prevention and mitigation decision-making and action. Technology transfer operation and training activities were conducted to increase competence within national and local co-operating agencies to independently maintain and operate an FDRS. Application education and demonstration projects aimed to increase capacity within central and local co-operating resource management agencies to understand and develop actions based on outputs of FDRS.

Adaptation Outcome

Expanded application of the FDRS in fire-prone areas of Southeast Asia.

- ☆ Seconded technical specialists from relevant government agencies to work at project field office in Jakarta.
- ☆ Established working relations with local counterparts at the Indonesian National Agency for Meteorology and Geophysics (BMG) and other relevant agencies.
- ☆ Commenced calibration studies for fire climate characterization, fuel characterization, and fuel mapping.
- ☆ Assessed adaptation needs of FDRS for Indonesia, Sabah, Sarawak, and ASEAN.
- ☆ Supported graduate student research relevant to FDRS calibration.

Operation Outcome

Enhanced awareness and capacity of regional networks to provide early warning for anticipated fires and transboundary haze before they become serious issues.

- ☆ Conducted introductory training and technical missions in fire science and FDRS concepts.
- ☆ Provided technical and administrative support to local agencies involved in drafting and promulgating new national fire control legislation.
- ☆ Formulated draft fire danger output products and discussed them with potential user agencies.
- ☆ Guided counterpart agency in its role as a facilitator among Indonesian agencies involved in FDRS operations and applications.

☆ Identified potential Indonesian FDRS operating agency and locations for provincial pilot initiatives.

☆ Formulated strategy for provincial pilot initiatives, which will complement newly established Fire-suppression Mobilization Planning in Riau and West Kalimantan.

Application Outcome

Enhancement of vegetation fire information and management systems in the region, to complement the FDRS.

☆ Supported development of Southeast Asia Fire Science Network.

☆ Actively contributed to regional ASEAN meetings related to fire and haze issues.

☆ Supported development and operation of ASEAN weather review committee.

☆ Presented FDRS technical information at several international conferences and meetings.

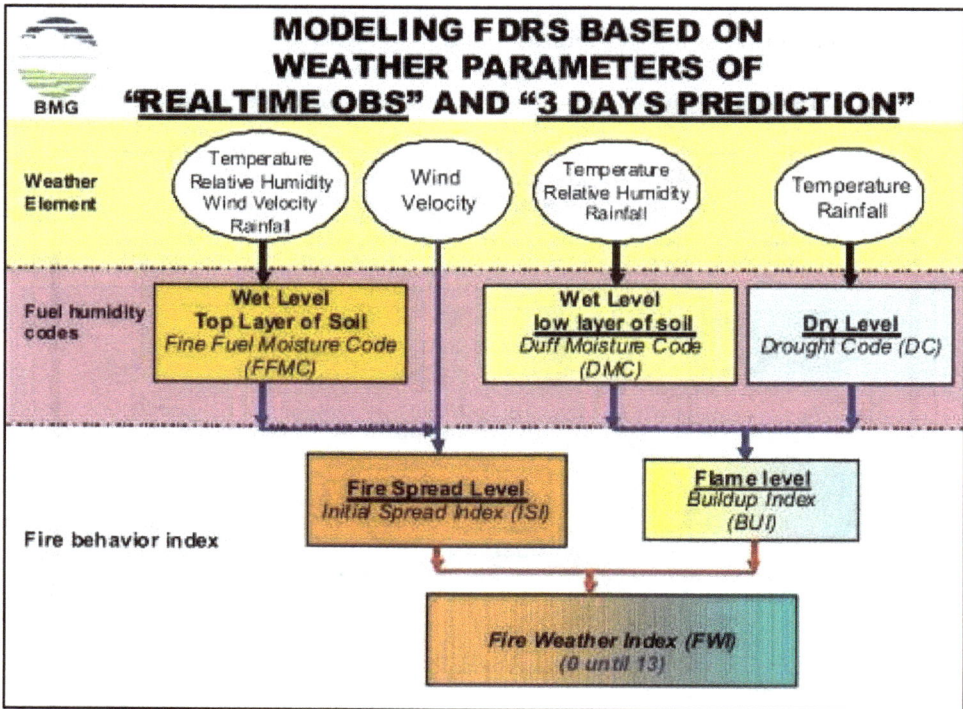

Figure 14.11: Modelling FDRS Based on Weather Parameters of Realtime Observation and 3 Days Prediction.

Indonesia Tsunami Warning System (InaTWS)

A reliable monitoring system of TWS relies on the performance of its sub system. These include the **monitoring system, analyses/ processing system, communication system, as well as dissemination system**. There is, however, a misconception related to this monitoring system. There is a tendency to assume that a fully automatic system will perform better then the manual one. Our experience clearly indicated that manual check and re-check in the form of interaction with the nearest station is a necessity for a reliable tsunami monitoring system. All seismic data are derived from sensors deployed throughout Indonesia. When earthquake occurs, the nearest sensors will receive the signal faster than the further ones.

Internatioanl Collaboration

Meteorological–Climatological Early Warning System (MC EWS)

Bilateral cooperation in MC EWS is very valuable, since assistance in the form of real time and forecast/ model output data is needed to be able to provide weather and climate forecast. WMO established GAW (Global Atmosphere Watch) in 1989 by merging GO3OS and BAPMoN, focusing on global networks for GHGs, ozone, UV, aerosols, selected reactive gases, and precipitation chemistry. GAW is a partnership involving contributors from 80 countries and it is coordinated by the Environment Division of WMO/AREP. Currently GAW coordinates activities and data from 24 Global stations, 200 Regional stations, and 19 Contributing stations. Indonesia is

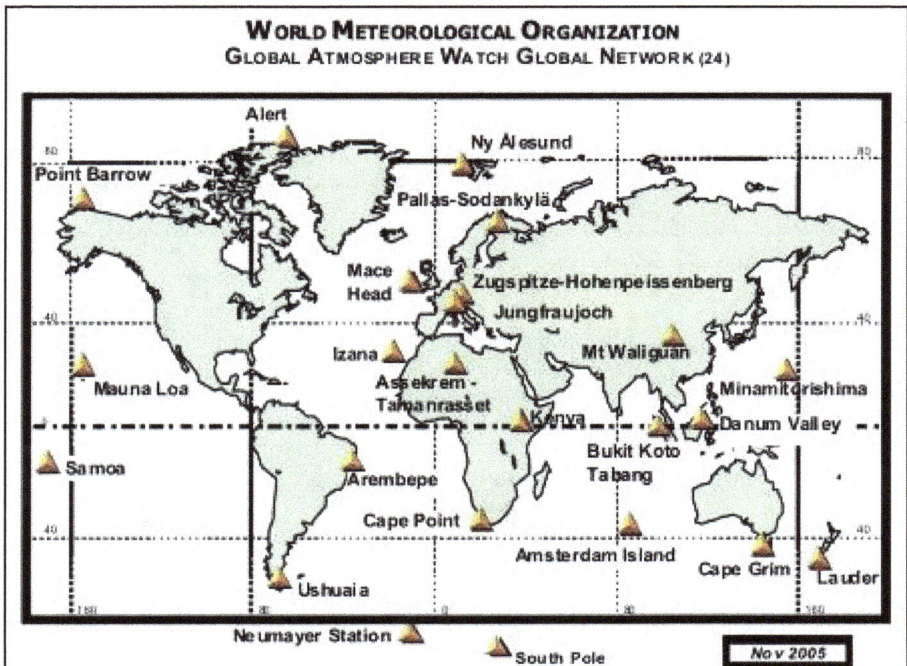

Figure 14.12: Global Atmosphere Watch (GAW) Network.

also operating Global Atmosphere Watch (GAW) Observatory in Kototabang – West Sumatra.

Tsunami Warning System (InaTWS)

Complexity and the extent of various aspects of the disaster which are beyond administrative boundary of Indonesia require extensive collaboration at international scale, bilaterally as well as multilaterally. Related to geophysical -induced disaster at present such international collaboration is carried out within the frame work of the IOC, CTBTO, WMO and ASEAN. Bilaterally, we should mention the collaboration with Germany, Japan, China, France, USA and Australia. The first four countries provide additional facilities such as seismic sensors and their processing system as well as assistances in building the relevant capacity. More seismograph stations will continue to be installed in the near future.

NATIONAL & DONOR COUNTRIES CONTRIBUTIONS

CONTRIES			2006		2007
			On Line	Stock	
Indonesia	108 seismo	454 accelero	44 (S) 44 (A)	6 (S) ---	---
China	10 seismo	10 accelero	---	10 (S)	10 (A)
Germany	21 seismo	21 accelero	6 (S) 15 (A)	6 (S) 15 (A)	9 (S) 21 (A)
Japan	15 seismo	15 accelero	15 (S)	15 (A)	
CTBTO	6 seismo	---	1 (S)	3 (S)	2 (S)
TOTAL	*160 seismo*	*500 accelero*			
France	3 TREMORS	upgrade	3 (T)	---	---
USA	Technical Assistance		OK		
UNESCO	Training		OK		
IOC	Expert		OK		

Figure 14.13: National and Donor Countries Contributions.

Advisory and Dissemination System

Dissemination Sub-System on Meteorological–Climatological Early Warning System (MC EWS)

As we all realize in most of developing countries, the level of knowledge on weather and climate is still quite poor. In collaboration with universities, BMG have developed Climate Field School programme.

The output of BMG basically is in the form of climate outlook/ prospect in scientific language (*e.g.* "Above Normal", "Below Normal"). This language is hard to understand by farmers. Universities will translate scientific language to the operational language. The operational language will be converted into the strategy of management (*e.g.* crop management strategy). A detailed example in the field of crop management strategy is illustrated below:

From BMG	→	Above Normal
University	→	in the form of mm/ quantitative
Office of District Level	→	strategy of crop management
Change the crop variety	→	*Adjust the plant time!*

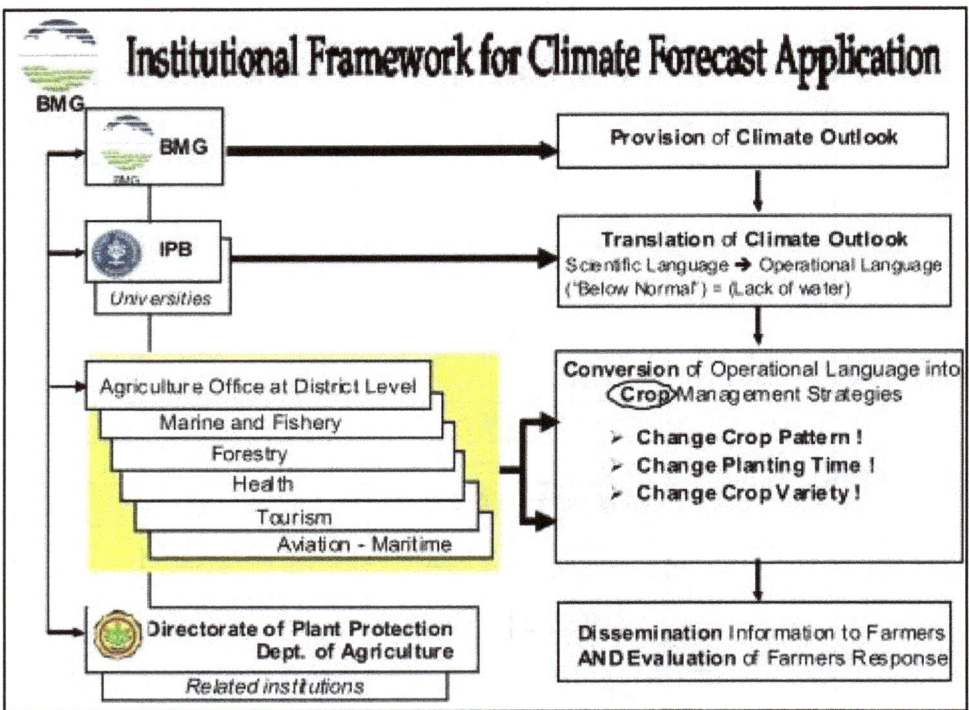

Figure 14.14: Institutional Framework for Climate Forecast Application.

Dissemination Sub-System on Tsunami Warning System (InaTWS)

When an earthquake occurs, the position of earthquake hypocentre, the initial rupture point of the earthquake and its magnitude are all unknown yet, we can identify when and where it occurred and the possibility of tsunami after the nearest sensors received and recorded the signal and send to the automatic processing machine in the headquarter for analyzing, processing and producing earthquake information. If the hypocentre is under the ocean and not too deep within the earth, and if its magnitude is sufficiently large, tsunami generation is possible. The Headquarter then issues a Tsunami Warning to the Regional Tsunami Warning Centre – to the National Police Headquarter – SMS cellular service – 11 TV Stations – Local Government. The initial broadcast informed the possibility of a tsunami but not yet confirmed. Concerning the confirmation of tsunami, DART buoy deployed in the ocean bottom around subduction zone, will measure the wave and send the data to BMG for further warnings.

The distance of the "fault zone" on the ocean bottom to the shore is on average only 250 km, meanwhile tsunami travel speed is around 800km/hour, so only less than 25 minutes the tsunami wave will arrive and hit a coastal area. That is to say that there's very limited time to give an official warning and people must make their own decision. Therefore it is very important that each person in school, village or town, know the warning signs and what to do as instructed by the district local government. In the case of Indonesia, we have only 20 minutes of golden time for evacuation.

Dissemination system is one of the crucial points for the success of the end-to-end Early Warning System. In the case of developing countries which consist of multitude number of remote small islands where communication system is relatively undeveloped, local administration structure and local mentally is rather complex, therefore dissemination system have to be specifically managed.

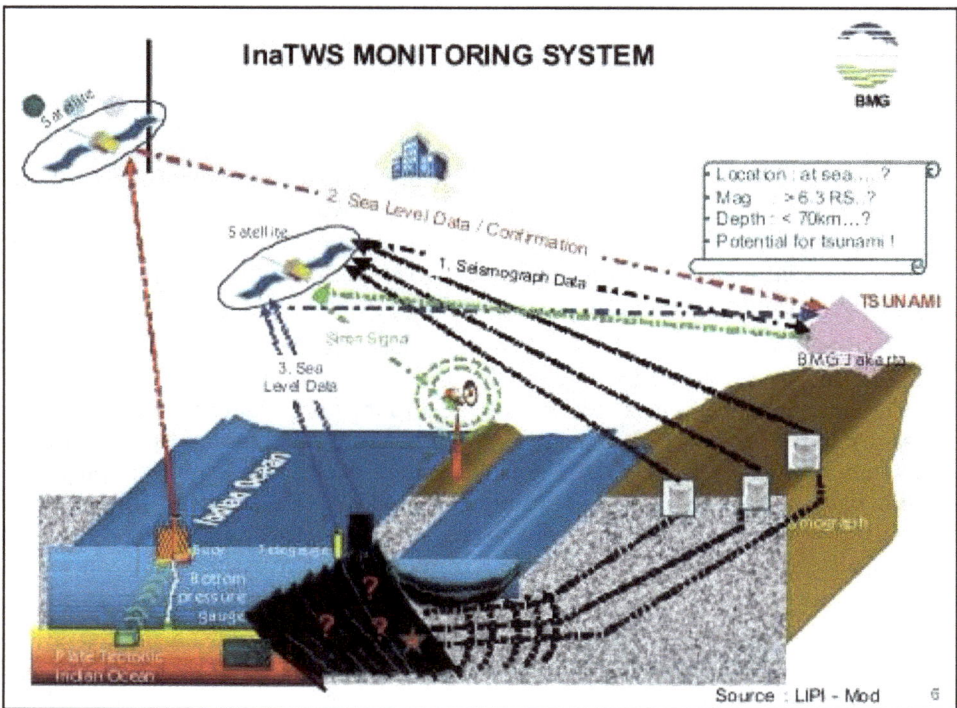

Figure 14.15: InaTWS Monitoring System.

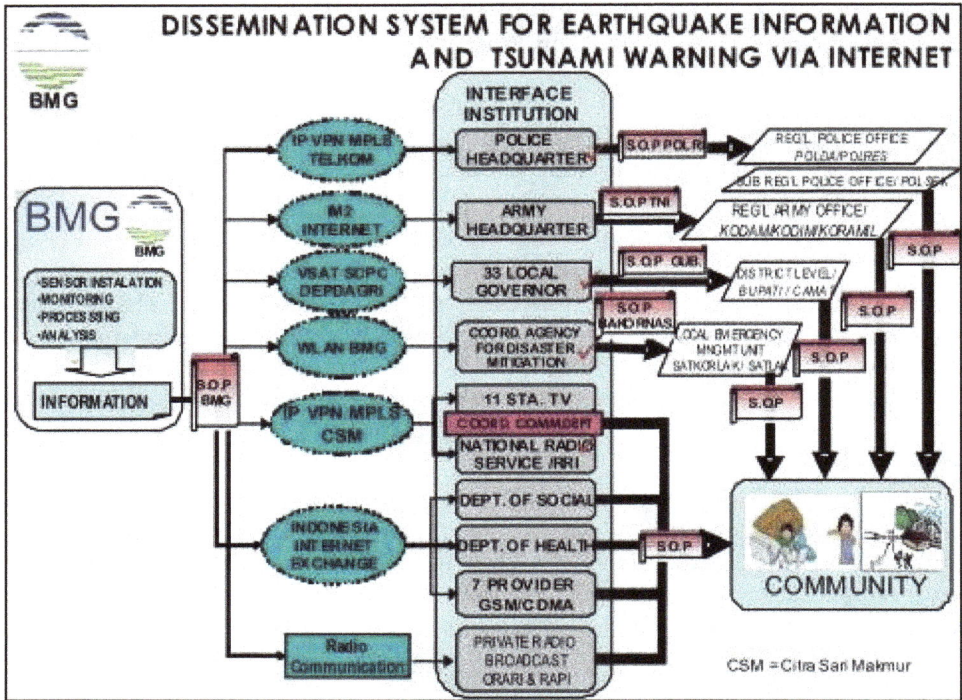

Figure 14.16: Dissemination System for Earthquake Information and Tsunami Warning Via Internet.

To disseminate risk information and warnings, there are Interface institutions that functioned as a bridge between BMG (as the early warning provider) and the end users. Interface institutions such as National Police Headquarter – Dept. Of Interior – Dept. of Social – Dept. of Health as well as TV – Radio Stations, issue warning information received automatically from BMG.

Beside via interface institutions, warning is also distributed directly via cellular phone to the related government agencies and departments, as well as municipal government. Currently, there are about 900 cellular phone receiving early warning messages. In the case of Aceh tsunami, the only communication link which worked was satellite – based hand phone. Recently, there exists a new innovation in the satellite based telecommunication system so called RANET, the radio along with a hand-sized satellite antenna which has strong reception capacity. It is able to receive abundant satellite audio and data programme, in most of Asia region via world space satellite AsiaStar.

Warning information to the public community in district and sub-district where telecommunication system is relatively weak is disseminated via RANET system. Considering the advantage of RANET capability, in collaboration with NOAA, BMG

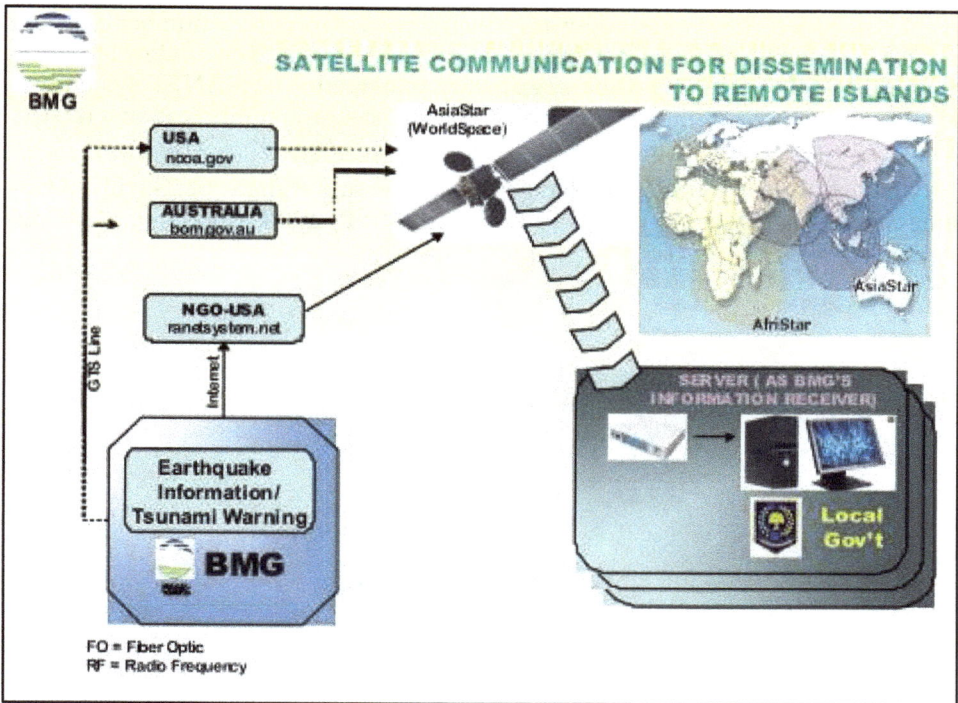

Figure 14.17: BMG Multi Mode Information Dissemination.

will set up RANET system in the office of sub-district government, where hazards might occur.

At present, internet-based receiving system consists of server and modem established in a number of provincial government offices, to which is ready to receive warning information from BMG Head Office.

Outlook

Recently, both meteorological disaster and geophysical disasters show a tendency to be intensified with series on anomalies.

Development of integrated meteorological observing network to acquire surface and upper air data, as well as utilization of reanalysis/ prediction data, raises the performance of BMG on providing weather, climate services as well as earthquake information and tsunami warnings.

Another aspect to be concerned is the dissemination system so the advisory and warnings received by the users on time. There are a number of interface institutions functioned as a bridge of the dissemination between BMG and the end users.

In the case of developing countries which consist of multitude number of remote small islands where terrestrial communication system is relatively undeveloped, the dissemination system has to be specifically managed. The RANET system is a breakthrough to overcome the frequently disturbances on communication terrestrial link.

In the future, to forecast the changes of weather and climate, BMG will develop state-of the art Numerical Weather Prediction (NWP) model and climate forecast technologies.

Chapter 15

Disasters and Solutions for Mitigating Natural Disasters in Vietnam

Phung Van Thanh
Department of Science and Technology,
Ministry of Science and Technology (MOST), Hanoi, Vietnam
E-mail: pvthanh@most.gov.vn

ABSTRACT

It is realized that confronting natural disaster is indeed a complicated matter, but the difficulties can be reduced if forecasting, warning system as well as preparedness are accurately and timely developed.

The efforts and steps taken by the Government of Vietnam are briefly described in this chapter. It also includes the type of disasters, policy measures and the recent development that the government has implemented.

Introduction

Vietnam is located in the South-East Asia and is one of the most disaster-prone countries in the world. The worst damaging disaster is flood, particularly when accompanied by typhoons. Typhoons raise sea levels and cause storm surges up estuaries, inundating valuable cropland. Typhoons destroy buildings due to their high-velocity winds, and generate waves which damage sea dykes protecting coastal land holdings.

The torrential rains accompanied by typhoons can cause flash-floods. I come without warning and regularly submerge low-lying areas. The runoff of these typhoons, when it added to monsoon rains swollen rivers, it creates floods which endanger river dykes and threaten to destroy millions of households. On the average, 8 to 10 typhoons reach Vietnam each year. Hundreds of people were killed. It is anticipated that the number of heavy storms and typhoons to hit Vietnam will increase both in number and intensity with global warming.

The river and coastal dyke systems in Vietnam are centuries old and suffer from piping, slides and local collapse from generation to generation. Vietnamese people always make efforts to strengthen dykes, to build flood protection structures, and to apply appropriate measures for disaster prevention in order to minimise disaster damage. The government ceaselessly pays attention to disaster mitigation, and co-ordinates with international organisations to increase co-operation in the field of disaster reduction aiming at guarding people well-being.

Disasters Management in Vietnam

The global changes of weather nowadays have caused increasing of disaster in both quantity and severity. Natural disaster mitigation is becoming a subject of profound global concern. The Government of Vietnam determines to carry out all measures that make them appropriate to the expectations of Vietnamese people, aiming at:

☆ Reduction of losses to human life,

☆ Reduction of losses to social property, and

☆ Reduction of losses caused by the standstill of production.

The measures cover both construction measures and non-construction measures. Construction measures are all but related to physical outputs such as dykes to control the flow of water. Along the coastal plains in the central part of Vietnam, people have built sea dykes to protect themselves against extreme events such as typhoons. These dykes are essential for agriculture in the central provinces. When these dykes are overtopped or braked, the fields are flooded by saltwater and rendered unproductive period for years. If this happens frequently, there are not sufficient resources in the community to repair or upgrade the dykes, and the people become impoverished and prone to malnutrition.

Non-construction measures are related with policies and regulation that supported rehabilitation process. For example, problem of population pressure in the in annually inundated floodway and in hazardous coastal areas have raised up and caused unauthorized settlement. In response to this, the Government has recently promulgated Statute on Dyke Management for Flood and Typhoon Mitigations setting out the responsibilities and power of the authorities to control development in flood and typhoons prone lands. In this regards, the government foresees that coordination beyond the construction measure related institutions is needed, such in specifying minimum flood level that it gives an alarm to the people lives within the potential disastrous areas to be evacuated. In that case, accurate and timely hydro-meteorological forecasts and warnings pose very effective tool.

Moreover, realizing the numbers of loss caused by the recent disasters in Chanchu and Xangsane, the government has taken further steps based on the assessment of the Central Committee for Flood and Storm Control (CCFSC) in rehabilitating people lives and their production. Those recommendations cover disaster preparedness, disaster response, and overcoming disaster consequences.

Accordingly, the government establish warning mechanism so-called the Warning Information System and National Flood and Storm Control are developed at every level of government, as follows:

At Central Level

The warning is issued by the Central Committee for Flood and Storm Control, through the Standing Office of Central Committee for Flood and Storm Control.

At Provincial Level

The Provincial People's Committee and Provincial Committee for Flood and Storm Control is responsible to issue warning through the Office of Provincial Flood and Storm Prevention.

At District Level

The District People's Committee and Provincial Committee for Flood and Storm Control disseminate the warning through the Standing Body assisted by the Provincial Committee for Flood and Storm Control.

At Village Level

Village People's Committee and Village Committee for Flood and Storm is responsible issuing the warning.

Based on the geographical and climatological features, as well as the disaster conditions, the Government of Vietnam have made decisive policy for each zone, as follows:

Northern Part of Vietnam

Strengthening dyke system, flood retardation and diversion structures to improve flood resistant coefficient of constructions and to protect essential population and economic areas against flood.

Central Part of Vietnam

Central Vietnam is narrow and topographically complicated, frequently affected by storms, and rapidly rainwater concentrated resulted in flooding. The decisive policy is to supplement active measures for flood prevention and mitigation, as well as for familiarising with floods.

Mekong River Delta

The decisive policy is to prepare measures for living with floods, to minimise damage caused by floods as well as to make use the advantages of floods for the sustainable development.

The approach to disaster response has also been shifted from defensive to offensive one. When disasters occur, the disaster response activities as well as the disaster damage assessment will be actively carried out and the disaster consequences will be rapidly overcome.

Based on experiences gained from the process of disaster prevention, response, and mitigation inside as well as outside the country, apparently the following lessons can be derived:

☆ Improve the gathering activity of disaster forecast information, timely organise the warning activity as well as provide particular measures to disaster-affected localities;

☆ Mobilise all forces from all social stratums for disaster prevention, particularly the armed forces considered as a core in all construction as well as in search and rescue activities. Mobilise the "mutual affection and love" spirit of the community to help disaster affected people in overcoming the disaster consequences;

☆ Regularly learn from experiences to promptly adjust and supplement appropriate measures as well as to strengthen steering machinery to obtain high leading efficiency.

☆ Strengthen the information system from central to local levels, strictly stipulate the regime of information dissemination to serve the disaster prevention and mitigation career;

☆ Enhance the community disaster awareness activity, diversify communication appearances so that all people can well-understand disasters and be able to respond to every situation of disaster;

☆ Extend research to all types of disasters to find out the appropriate measures to minimise disaster damage, and to improve the efficiency of the guideline of "four on-the-spots": directions, forces, materials, and logistics on-the-spots, particularly in rehabilitating flood and storm control structures, to mobilise the role of the military force and the close direction of local authorities; and

☆ Develop strategy for disaster mitigation for the whole country, based on each disaster zone.

Conclusion

It's very difficult to confront natural disaster. The improvement of accuracy and timely forecasting and warning systems as well as preparedness can reduce the damage to minimum levels.

Chapter 16

Progress in Implementation of Disaster Management Programme in the Union of Myanmar

Aye Ni Aung
Department of Technical and Vocational Education,
Ministry of Science and Technology, Myanmar
E-mail: ayeniaung@z6.com

ABSTRACT

Myanmar faces annually 70 per cent of disasters caused by flood, 13 per cent by storm, 10 per cent by flood and 7 per cent by other cause. However, the severest disaster that caused lives was tsunami in 2004 where Myanmar suffers 61 lives and 42 injures.

Accordingly, Myanmar has stepped up its disaster management activities after that. Presently, Myanmar has its own system and practice for disaster prevention and preparedness. Accordingly, the government has initiated new laws at national level for disaster risk reduction management in the new constitution.

This chapter explains briefly systematic operational implementation related with the disaster management system that has been carried out by the government of Myanmar.

Introduction

Myanmar is located in Southeast Asia, between latitudes 09°32'N and 28°31'N as well as longitudes 92°10'E and 101°11'E, and encircled by Laos and Thailand in the east and southeast as well as China in the north sides. The country covers an area of 677,000 square kilometers (261,228 square miles) ranging 936 kilometers (581 miles) from east to west and 2,051 kilometers (1,275 miles) from north to south. Over

50 per cent of the total land area is covered with forests. The coastline is about 2400 km length and covers almost all east coast of Bay of Bengal.

There are wide varieties of natural hazardous phenomena in Myanmar, including earthquakes, volcanic activities, landslides, tropical cyclones and other severe storms, tornadoes and high winds, river floods and costal flooding, wildfires and associated haze, drought, sand/dust storms, insect infestations. Annually, Myanmar has to suffer from impacts of disaster: fire, storm, flood, and earthquake. 70 per cent of disasters are caused by fire, 13 per cent by storm, 10 per cent by flood and the remaining 7 per cent by other causes of disaster. The severest natural disaster in Myanmar is the earthquake. Myanmar was affected by tsunami of 26 December 2004 and it left 61 death and 42 injured. Myanmar stepped up disaster management activities management activities with momentum after that.

Today, to manage natural disaster efficiently is one of the most challenging problems to the world, none-the-less in Myanmar. In the last two decades, over 3 millions people have died in natural disaster. This challenging problem is calling for the comprehensive efforts of the world. During last decades the world has been the witness of the numerous natural disasters. This report briefly explains the natural disaster management system established in Myanmar.

Disaster Management System

Myanmar has its own system and practice for disaster prevention and preparedness base on its own social, economical, cultural and administrative practices. In order to carry out disaster preventive measure effectively, the Central Committee for Natural Disaster Prevention, Relief and Resettlement has been formed by the Guideline of State Peace and Development Council's Security and Management Committee. The purposes of the committee, among others, are:

☆ To involve, integrate and co-ordinate the input of different organizations; and

☆ To provide a comprehensive, systematic approval of disaster management.

The chairman of the committee is Minister for Social Welfare Relief and Resettlement. The members are heads of Departments concerned. State, Divisional and Township level committees are also established through which disaster prevention activities are being implemented.

The Government also established a Central Committee for Disaster Prevention and Relief. The objective is to carry out effectively disaster preparedness and prevention measure. This Committee is a policy formulating body headed by the Minister of Home and Religious Affairs. Under the aegis of this committee, a National Disaster Prevention Relief and Rehabilitation Committee was established to facilitate practical implementation of the preparedness and preventive measures. This committee is headed by Deputy Minister of the Ministry of Social Welfare with 9 members from other Ministries

Forecasting and Warning

The installation of the Automatic Picture Transmission (APT) System in Kaba-Aye in 1973 as well as completed by the 10 cm WSR 74S Radar at Kayukpyu in 1979 have enhanced our capability to survey and track down tropical cyclone effectively. The APT results satellite imageries that facilitates the Department of Meteorology and Hydrology (MHD) to issue storm warning timely. Myanmar Television and Broadcasting Stations disseminate the warning to public.

The Department had also developed an analytical and empirical prediction model for Storm Surge during 1980. In 1986, the prediction model was modified to predict not only the maximum surge height but also the surge envelope for the respective land fall point along Myanmar coast. However, in the absence of the direct observations of the storm parameters, reliable source of data and better computing facilities, the method used recently in DMH provide a valuable first guess for the prediction of Storm Surge within acceptable accuracy.

Under the guidance of the Central Committee for Disaster Preparedness and Prevention the following counter-measures against Tropical Cyclone hazards are practiced in the country:

1. Taking precautionary measures during warning period, such as boarding-up, buildings, closing public facilities;
2. Moving of people to safe shelters and providing necessary support;
3. Providing public awareness programmes.

Floods

The Department of Meteorology and Hydrology is using advance techniques on a microcomputer based River forecasting and Flood warning system for issuing Flood warning and bulletin to the users and public. The warning time for issuing flood warning is from one day to 12 days in advance. The dissemination of forecasts and warnings is done through SSB transceivers, telephone, Myanmar Radio and Television Service.

Mitigation of flood damage is undertaken in a number of places by constructing embankments and dams in flood prone area. The following countermeasures against flood disasters are practiced in the country.

1. Issuing early flood warnings in order to carry out flood control arrangement;
2. Laying out land-use regulations and building regulations for reducing flood hazards;
3. Preparing evacuation plans and arrangements;
4. Providing emergency equipment facilities and materials, such as special boats, sand bags as well as designated volunteers for implementation of emergency measures;
5. Providing public awareness programmes.

Earthquakes

Myanmar Naing Ngan situates on the Aliped -Himalayan earthquake belt where devastating earthquakes had occurred from time to time. Counter-measure against these disastrous events is a real necessity. The Department of Meteorology and Hydrology is the sole department which is responsible for earthquake observations. Seismological news is broadcasted and published to the people by the DMH with the help of Myanmar News Agency.

Table 16.1: List of Strong Earthquakes in Myanmar.

Sl.No.	Date	Epicenter		Magnitude (Richter Scale)	Remarks
		Latitude (°N)	Longitude (°E)		
1.	23-05-1912	21.00	97.00	8.0	North of Taunggyi, serious landslide.
2.	06-03-1913	17.00	96.50	7.0	"Hti" of Shwe-maw-daw Pagoda grounded.
3.	05-07-1917	17.00	96.50	7.0	"Hti" of Shwe-maw-daw Pagoda grounded.
4.	19-01-1929	25.90	98.50	7.0	Brick buildings destroyed at Htaw-Gaw.
5.	08-08-1929	19.25	96.25	7.0	Railway lines destroyed at Swa.
6.	16-12-1929	25.90.	98.50	7.0	Landslide at Htaw-Gaw.
7.	05-05-1930	17.00	96.50	7.3	Many houses destroyed, 500 killed in Bago. Some houses destroyed, 50 killed in Yangon.
8.	03-12-1930	18.00	96.50	7.3	Some houses destroyed, about 30 killed in Phyu.
9.	27-01-1931	25.60	96.80	7.3	Brick buildings collapse, landslide at Karmine.
10.	12-09-1946	23.50	96.00	7.5	Pagoda collapse at Ta-gaung.
11.	15-08-1950	28.50	96.50	8.6	Under the influence of Assam earthquake, Chindwin river at Mawlaik and Kalewa, Ayeyarwady river at Aunglan flow upstream.
12.	16-07-1956	22.00	96.00	7.0	Pagoda and buildings at Sagaing destroyed, about 40 killed. Sagaing bridge moved slightly.
13.	08-07-1975	21.50	94.70	6.8	Many historical pagodas destroyed 2 killed near Bagan.
14	05-01-1991	23.48	95.98	7.1	Landslide and some buildings destroyed at Tagaung, Hti-gaint, Kawlin and Thabeikyin and surrounding area. 2 killed.
15.	22-09-2003	19.94	95.72	6.7	Landslide, liquefaction and sand Eruption. Pagodas, some bridges, houses and schools destroyed at Taungdwingyi and surrounding areas. 7 killed.

The most active earthquake observations are conducted in Sagaing division where Sagaing fault is the main cause of producing tremors. The strong Earthquakes in Myanmar is shown below.

Under the direction of the Central Committee for Disaster Preparedness and Prevention, the following countermeasures against earthquake hazards are practiced in the country.

1. Designating and constructing buildings or structures, such as that will withstand the maximum possible vibration of earthquakes and laying out building regulations;
2. Instructing a property against earthquake damage and at the same time to lay out Land-use regulations;
3. Educating the people about the earthquake phenomenon and the countermeasures against earthquake; and
4. Training and research works with emphasis on formulation of a prediction technique for earthquake based on the recordings, study of the occurrence and variation of geophysical phenomena.

Droughts

Drought can be considered mainly as meteorological, hydrological and agricultural droughts. According to the WMO's definitions, meteorological drought can be defined as prolonged absence or poor distribution of precipitation less than 60 per cent for two years continuously in any large areas in Myanmar.

Regarding hydrological drought, a study on drought identification is recently made in the Department using monthly rainfall and runoff data. The method is based on a modified method suggested by Mohan and Rangacharya (1991). The characteristics of drought used in identification are on-set and termination of drought, drought duration and performed well in the identification and characterization of drought using monthly rainfall and runoff data.

Disaster Management Plan of Myanmar

Department of relief and resettlement have cooperated with the Department of Health, Department of Meteorology and Hydrology, Fire Services Department, Human Settlement and Household Development Department, Irrigation Department and Myanmar Red Cross society in the field of disaster prevention and reduction measure.

Fire Prevention

The State Peace and Development Council is the authority concerned to manage a systematic resettlement of homeless who trespassed in restricted areas to new towns. On the other hand, the authority made town plans at the fire stricken areas as post-disaster activities. The victims were provided rehabilitation plans, resettlement and development schemes.

Fire Services Department under Ministry of Social Welfare Relief and Resettlement is responsible for fire precaution, fire prevention, extinction, training of fireman, relief and rescue work, educating the public for awareness of fire and disasters.

Reducing vulnerabilities can be accomplished by physical plan and construction. Some of the examples are Meiktila Fire (1991) and Myangyan Fire (1993). As the main causes of fire outbreak are due to unplanned development and use of flammable construction materials, the authorities have undertaken the task of physical planning to prevent potential disaster and have also established low cost building materials with indigenous raw materials. Systematic establishing of new towns in accordance with town plan is another activity to reduce Fire disaster.

Cyclonic Storm and Flood Preparedness

Having a long coastal line along the western part of the country, Bay of Bengal is regarded as cyclonic vulnerable area. Being a heavy rainfall country, Myanmar suffers from flood disaster in mid-monsoon period of August to October. Accordingly, in the cyclonic prone area, embankments which consists of refuge shelters and drinking water ponds were constructed such as in Pauk-taw, Myebon and Minya Township in Rakhine state. During the cyclone season (April, May, September, October), local people are led to these high mounds and shelters in case of emergency to avoid from storm surge and strong wind.

With regards to that, Department of Meteorology and Hydrology is responsible for improving cyclone and flood warning and forecasting system. Hence, broad dissemination of warning are being made through the mass media such as television, radio, wireless, and newspaper.

In the delta regions, where flooding of the river is the problem, the dykes and water barriers are maintained and reinforced as necessary by Irrigation Department. In area of Rakhine state which are vulnerable to cyclone and storm surge, earthen mounds are constructed.

Preparedness and Other Operational Measures

Since it is realized that relief measures and post disaster action are by themselves not sufficient, emphasis is also given to planning and prevention. The need to make people familiar with and understand the natural disaster prevention measures is becoming priority. Short term training courses in the states and divisions annually will be continued for the training to trainers who in turn will organize and conduct the training their own regions with a view to achieve community awareness.

As soon as cyclone warning together with storm surge warning or flood warning is issued through radio and television, the local authorities start implementing the preparedness measures, such as: alarming local people to take precautionary measures, reinforcing embankments with sand bags, providing life guards and security guards at flood prone areas, providing vehicles or motorized flood boats if necessary and providing foods, medicines, blankets and shelters for the victims.

Such operational measures are usually carried out at the township or district hit by any natural disaster, and at the same time the responsible personnel from the National Disaster Prevention, Relief and Rehabilitation Committee come to the site immediately in order to give necessary instructions.

To carry out works systematically and to manage immediately in the disaster areas throughout the country, the Five Priorities Action are being implemented:

1. To ensure that disaster risk reduction is a national and local priority with a strong institutional basis for implementation,
2. To identify assess and monitor disaster risk and enhance early warning,
3. To use knowledge, innovation and education to build a culture of safety and resilience at all levels,
4. To reduce underlying risk factor, and
5. To strengthen disaster preparedness for effective response for all levels.

Conclusion

Myanmar is trying its utmost effort in order to carry out effective disaster management programme. The government is initiating to formulate new laws at national level for disaster risk reduction management in the new constitution. Being a developing country, Myanmar naturally suffers from completeness in managing the disaster systematically.

However, with mutual understanding posed by regional cooperation and cooperation with UN as well as other international bodies, Myanmar strongly believe that not only Myanmar but also every country in our society will be benefited in upgrading the services in the field of Disaster Management.

Vellore Resolution 2007

Regional and International Cooperation in Disaster Mitigation and Management under the aegis of the Centre for Science & Technology of the Non-aligned and Other Developing Countries (NAM S&T Centre)

We, the representatives of the non-aligned and other developing countries from Bangladesh, Brunei Darussalam, Colombia, India, Indonesia, Malaysia, Mauritius, Myanmar, Pakistan, South Africa, Sri Lanka, United Arab Emirates, and Vietnam.

Thank

☆ The Centre for Science and Technology of the Non-aligned and Other Developing Countries and the Centre for Disaster Mitigation and Management (CDMM) of the VIT University, Vellore, the joint hosts of the International Roundtable Meeting on 'Lessons from Natural Disasters-Policy Issues and Mitigation Strategies' held at Vellore, India during 8-12 January, 2007.

☆ Our respective governments and sponsors who have made our participation at this very important meeting possible.

And

☆ Place on record our appreciation to the Centre for Disaster Mitigation and Management at the VIT University for providing the interactive platform, excellent ambience for the meeting, fine arrangements and kind hospitality.

Acknowledge with Gratitude

☆ The condensed wisdom emanating from the mammoth efforts made by the United Nations, International Strategy for Disaster Reduction, World Governments, Knowledge-based institutions across the globe, voluntary agencies, corporate sector, community-based organizations and others, reflected in the Earth Summit (1992), Yokohama Strategy (1994), Millennium Declaration (2000), World Summit on Sustainable Development (2002) and Hyogo Framework for Action (2005-15).

☆ The initiatives taken and efforts made by different countries to implement policies and programmes to prevent and mitigate disasters.

☆ Global investments and achievements made so far in harnessing the power of science and technology to mitigate and manage disasters, especially developing the power of geo-informatics (Remote Sensing, Geographic Information Systems and Global Positioning Systems), integrated communication technologies, climate change studies and weather forecasting.

Note with Concern that

☆ In many developing countries disasters are not yet being looked upon as opportunities to 'build back better', and efforts and investments to learn from disasters leave much to be desired.

☆ The culture of learning from disasters is growing at a very slow pace. Although we cannot recreate the lost lives, we can certainly endeavor to ensure that no more lives will be lost on recurrence of such events in future by making sincere and determined efforts to learn from disasters on a global level.

☆ Unplanned and non-engineered constructions, well known to be the massive killers (which give natural hazards a bad name), continue to be built in many developing countries for a variety of reasons including slack financial and techno-legal regimes, out-moded building byelaws and corrupt practices.

☆ Views of the first responders, eyewitnesses, survivors, crisis managers, policy makers, meteorologists, oceanographers, seismologists, geoscientists, planners, builders, researchers, educators, social scientists and NGOs are seldom systematically studied, analyzed and documented to find appropriate solutions as mitigation strategy.

☆ Proactive initiatives are seldom taken to generate reliable and continuous field data to bridge the information gaps and make practical use of the new knowledge, procedures and techniques to improve disaster specific investigations, measurements, mapping, analyses and design.

☆ People need right information at the right place at the right time, as much as they need timely food, water and medicine. Not enough is being done to provide easy access to useable information.

☆ The pooling of the brains, the building of the synergies, leveraging the capacities and advancing the frontiers of research by fostering, promoting and sustaining regional and international cooperation, especially among and between the NAM member countries has not yet been tapped, despite the fact that the tsunami of 26 December, 2004 triggered from Sumatra devastated coastal communities as distant as north-eastern Africa, and the tsunami triggered by the 1960 Chile earthquake devastated regions as far away as Japan.

☆ Many of the NAM and other developing countries are hit by recurring natural disasters and their problems are more or less of the same genre, and there remains great scope to build joint programmes and partnerships in science and technology in disaster mitigation and management. Since most of the developing countries lack even the minimal of resources and wherewithal to fight natural disasters, they may welcome regional and international cooperation.

Urge with Great Hope and Expectation that

☆ Policy makers will look at the disasters beyond the rebuilding process, integrate disaster management with the development process and avail every opportunity to test their policies against disasters.

☆ When planning for immediate action, disaster management machinaries should be swift enough to respond to complex situations unfolded by every type of disaster. While planning for on the scale of a decade, the endeavour should be aimed to enhance safety through a systematic planning and implementation of a time-bound programme. However, for long-term planning, a determined effort should be made to learn from disasters and spread the culture of safety.

☆ Urban planners, architects and engineers should avail of every opportunity offered by the disasters to find out what really went wrong. They should reveal, and not conceal, the failure stories.

☆ Disaster managers should introspect on what ails the disaster management machinery and renew commitment to revamp the disaster mitigation and management system.

☆ Financial, legal, and enforcement authorities and institutions should continuously reshape and reinforce techno-legal regimes to give teeth to the implementation of the Disaster Mitigation and Management Plan.

☆ Community based organizations and communities should gradually move to front-line positions and enhance in-house capabilities.

☆ Individual households should realize that everything in life can be lost in a fraction of a second and the impact of which can travel through generations, if they continue to defy nature and refuse to learn from the recurring disasters.

☆ Scientists should foster, promote and sustain the culture of scientific investigation, multi-disciplinary research and development concerning all aspects of understanding natural hazards for managing natural disasters. They should attempt to rigorously test and validate their theories and look for fresh ideas to re-write their research proposals in search of cost effective and speedy solutions. The learning process never ends, the paradox being that the more we learn, the more remains to be learned.

☆ Countries should develop short- and long-term Rolling Strategic Plans in view of the fast emerging scenario of multi-hazards and ensure their implementation through dedicated multi-disciplinary expert groups and prepared communities, modernization of investigational tools, planned production of large-scale base maps, systematic investigation and research, effective execution of sound land use planning practices, and spread of disaster education and training.

☆ The pace of scientific studies on diverse aspects of natural disasters, their mitigation and early warning should be accelerated and should form an integral part of disaster management.

☆ Investments on disaster risk reduction strategies should be enhanced and risk distribution through the insurance sector should be encouraged. Incentives/subsidies for adopting 'disaster resistant' features may be considered.

☆ Higher orders of investments should be made in documenting disaster and in developing manuals, guidelines and disaster education aids such as the knowledge based products, for example, the CD ROMS on disaster produced by CDMM at the VIT University.

Recommend that

1. The countries that have accrued the wisdom of integrating disaster management within their national development processes should demonstrate it by sharing concrete examples for others to emulate.

2. Every country should invest appropriately and adequately in reliably mapping and forecasting multi-hazard scenarios. Time-bound action plans backed by adequate resources should be evolved for the regions known to be most vulnerable.

3. Special attention should be paid to protect and preserve cultural and natural heritage such as archaeological monuments, bio-diversity parks etc., endangered by natural disasters.

4. Adequate attention and investments are also necessary towards the protection of strategic and life-line buildings, roads and communication systems.

5. Individual professionals, study groups and organizations must exercise utmost restraint in publishing hazard maps for public use without adequate test of reliability by field validation. Government-funded organizations should not allow inclusion of the unvalidated maps in any of their official reports and publications, as such maps represent no more than the work in progress.

6. Greater levels of investment are necessary for studies and projects leading to forecasting, prediction and early warning of different types of disasters. Higher priority and greater investment than at present are also essential to create good examples of functional, reliable and timely early warning systems against disasters.

Recommendation for Immediate Implementation

1. The participating delegates unanimously agree to launch an e-Newsletter titled 'Natural Disasters and Development' highlighting natural disasters in the non-aligned and other developing countries. It was agreed that all the delegates will serve as focal points for their respective countries, and the Centre for Disaster Mitigation and Management (CDMM), VIT University, India would assume the coordinating role and issue the e-Newsletter to all members once every two months, to start with. Gradual value addition to the e-Newsletter may be in the form of Readers Forum and dissemination of disaster-related information. The next step would be to create a Disaster Knowledge Network for the benefit of developing countries.

2. A compendium titled 'Natural Disaster Profiles of Participating NAM Countries' prepared by CDMM would be appropriately modified and enlarged to provide holistic coverage with the inputs provided by the participating countries. The final compendium will be disseminated by the NAM S&T Centre.

3. A discussion paper titled 'A Profile of Disaster Risk to Cultural Heritage Sites in Participating NAM Countries' prepared by CDMM would be suitably revised and enlarged. Participating delegates will provide their respective inputs to CDMM by 31st March 2007. The draft document will then be circulated for validation and the final compendium will be submitted to the NAM S&T Centre by 30th September, 2007 for further dissemination to the NAM and other developing countries.

4. NAM S&T Centre agreed to hold a meeting to develop the above compendium into a project through interactive dialogue, should any of its Member States and other developing countries agree to host the meeting. In this context, the Director, NAM S&T Centre requested the representative from South Africa to explore the possibility through the African Regional Office of ICSU, the Department of Science and Technology and other concerned agencies in South Africa.

5. An Association of Natural Disaster Experts from the NAM and other developing countries will be established to network the experts and to foster, promote, and sustain partnerships among the developing countries. The first step in this direction will be to compile a directory of experts on natural disasters. The participating delegates agreed to provide their inputs digitally to the CDMM, which in turn agreed to finalize the directory and make it available to the NAM S&T Centre for its further dissemination.

6. Capacity building is a common concern, especially of all the NAM and other developing countries in so far as disaster management is concerned. Information on research, education and training facilities available in the member countries will be exchanged through the envisaged e-Newsletter to pave the way for bilateral and multi-lateral regional and international cooperation.

7. The NAM S&T Centre will bring out a publication on the deliberations of the International Roundtable Meeting edited by Dr. R.K. Bhandari, Chairman, CDMM. Efforts will be made to mobilize the required resources.

8. It was agreed to explore funding possibilities, not only to kick start the above activities, but also to sustain, enlarge, and add value to them.